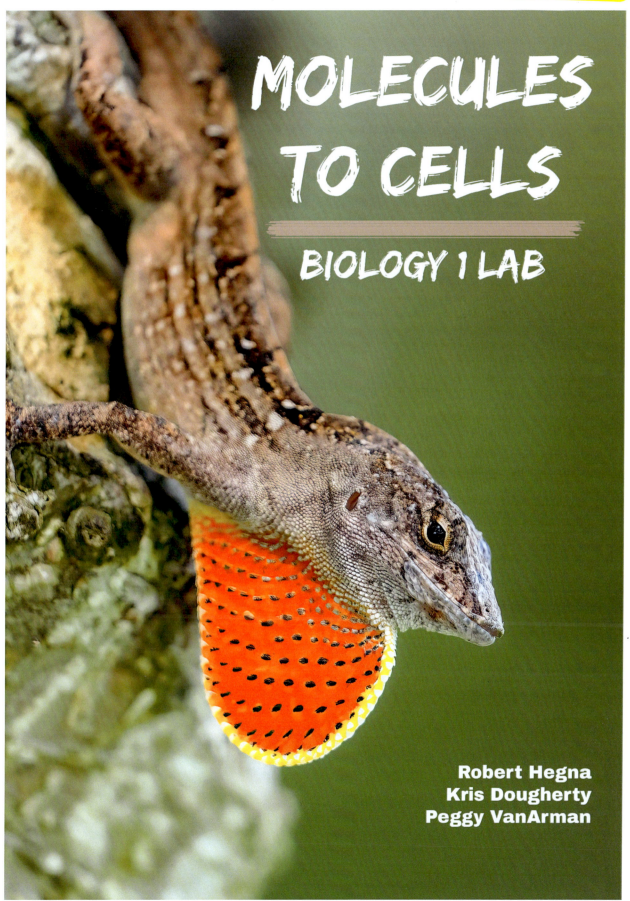

MOLECULES TO CELLS

BIOLOGY 1 LAB

Robert Hegna
Kris Dougherty
Peggy VanArman

Blue Frog

Molecules to Cells: Biology 1 Lab

ISBN-13: 979-8-218-45112-7

© Chapter 2, 3, 5, 6, 8, and 9 header images by Robert Hegna.

© Chapter 4 header image by Peggy VanArman.

Chapter 1 header image is licensed from Adobe Stock

Chapter 7 and 10 header images are licensed from Shutterstock.com

Printed in the United States of America. 10 9 8 7 6 5 4 3 2 1

Table of Contents

Chapter 1 1
Safety and Metrics

Chapter 2 19
Scientific Research

Chapter 3 35
Macromolecules and Cell Chemistry

Chapter 4 49
Microscopes

Chapter 5 73
Cell Membranes

Chapter 6 89
Enzymes

Chapter 7 103
Photosynthesis and Respiration

Chapter 8 119
Cell Division

Chapter 9 139
Heredity and Genetics

Chapter 10 157
Molecular Genetics

Important Notes to the Student

About this book:

Biology lab classes are where the rubber meets the road, or more aptly, where the pipette meets the test tube! They are designed to reinforce the concepts you learn in lectures and to teach you techniques used in a wide array of jobs and future studies you may undertake as you advance in your career. This book is your trusty guide to help you succeed in your biology lab class, offering more than just a list of procedures.

- Each lab includes important background information to help you prepare ahead of time and to serve as a handy reference during class.

- Let's face it, with all the instructions about techniques and procedures, it could be easy to skip past a critical task. But don't worry – we've got you covered! To keep you on track, activities that need to be completed during class are marked with a **microscope icon**, ensuring you don't miss any critical tasks while juggling multiple activities (which you often do in lab classes).

- Post-lab questions are included to help you reflect on the activities and experiments you perform, reinforcing your understanding and ensuring you get the most out of each lab session.

How to Be Successful in Lab:

- **Come Prepared** - Read the lab chapter and any assigned materials before class to understand the objectives and procedures. This preparation helps you make the most of your limited lab time.

- **Attend Every Lab Session** - Labs are essential for hands-on learning and mastering techniques. Missing labs means missing out on critical skills and experiences that can't be gained from just reading notes.

- **Actively Engage** - Participate fully in all lab activities, ask questions, and discuss findings with your instructor and peers. Active participation not only enhances your understanding but also makes the lab experience more enjoyable.

- **Review and Reflect Regularly** - Constantly review lab materials to prepare for exams. Learn key terms, concepts, and familiarize yourself with the equipment and its parts. Reflect on the experiments and activities you performed to understand the reasons behind them and the results observed. Complete the post-lab questions.

- **Utilize Resources and Manage Time Efficiently** - Take advantage of office hours, tutors, and online resources. Be prepared, so you can manage your time effectively to complete tasks within the lab period. This also makes sure you have time in lab to ask your instructor for help in understanding background or results.

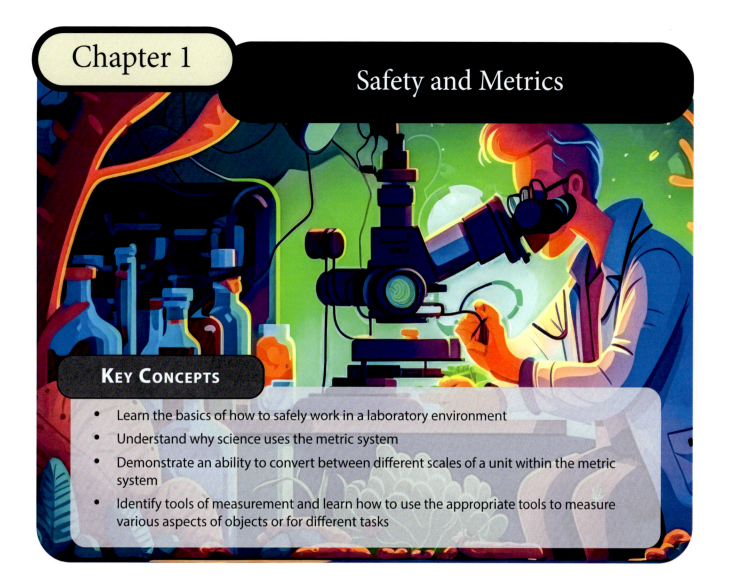

KEY CONCEPTS

- Learn the basics of how to safely work in a laboratory environment
- Understand why science uses the metric system
- Demonstrate an ability to convert between different scales of a unit within the metric system
- Identify tools of measurement and learn how to use the appropriate tools to measure various aspects of objects or for different tasks

Introduction

Welcome to your first college biology lab class! Throughout this semester, you will dive into the hands-on aspects of biology. In Biology 1, you'll learn to work in a lab environment, performing activities and experiments that reinforce your lecture material. Mastering two fundamental skill sets is crucial: proficiently using various tools for measurement and ensuring lab safety. One of the key characteristics of science is it's quantitative nature, so learning proper methods to measure something is absolutely essential for any scientist. Of course, knowing how to avoid blowing yourself up in a lab is just about as important. Don't hesitate to ask questions and seek help at any time during the semeter—this is a learning environment designed to support your growth and curiosity.

1.1 Safety in the laboratory

Lab work is done in nearly every sub-discipline of biology and in nearly every career that requires biological training. Therefore, learning how to work safely in a lab environment is one of the key objectives of this course. Practicing safety techniques in a lab setting is crucial to prevent accidents and ensure the well-being of everyone involved. Laboratory environments often involve the use of hazardous chemicals, biological materials, and sophisticated equipment, all of which pose significant risks if not handled properly.

In addition to safeguarding individuals, practicing lab safety is essential for maintaining the integrity of scientific research. Contaminations, inaccuracies, and equipment damage resulting from unsafe practices can compromise experimental results, leading to false conclusions and wasted resources. By following established safety procedures, researchers can ensure that their work remains credible and reproducible. This not only upholds the quality of scientific investigation but also fosters trust and collaboration within the scientific community.

Always be aware of your surroundings and remember that you are in a lab. To minimize the risk of injury, consistently conduct yourself with an acute awareness of your environment. Wearing proper attire is fundamental: a lab coat, long pants, and closed-toed shoes protect against spills and falling objects. Additionally, gloves and protective eyewear are essential when handling hazardous materials to prevent skin contact and eye injuries.

If a safety issue or emergency arises it is also important to know what equipment is standard in laboratories and where it is located. Equipment used in accidents or emergencies includes fire extinguishers, first aid kits, and bodily fluids clean-up kits. There are also more specialized items, such as eye wash stations for flushing your eyes if you get a chemical in them, and showers for when large amounts of a harmful liquid spill on a person's clothing and body.

Here are some commonplace specific rules that pertain to nearly any laboratory environment:

- Attire must be appropriate for a lab setting. Students must wear closed-toed shoes, and no hats, scarves, sunglasses, headgear, earbuds, or headphones are allowed. Hair that is shoulder length or longer must be tied up to prevent it from getting into chemicals or tangled in equipment.

- Lab coats should always be worn as an additional layer of protection.

- No food, drink, gum, water, or medicines are allowed in lab rooms. Additionally, no makeup, lip balm, contact lens solution, or artificial tear/wetting fluids may be used in the lab.

- If a student is injured during the lab (e.g., cuts, slips, trips, falls, chemical spills), notify the instructor immediately for medical assistance, spill clean-up, and incident reporting. Accidents involving bleeding, open wounds, or vomiting must be treated according to OSHA guidelines for the clean-up and disposal of blood-contaminated items.

- If chemicals or other substances get into a person's eyes, use the eye wash station for at least 15 minutes. If chemicals or toxic substances contact the skin, wash the affected area or the entire body in the shower for at least 15 minutes. In cases involving strong acids or bases, clothing may need to be removed, and the classroom will be cleared to ensure privacy and modesty.

- Notify your instructor of any broken glass; do not clean it up yourself. Use a piece of cardboard or several layers of paper towels to pick up glass, and place the pieces in the designated broken glass container. Similarly, notify the instructor of any broken scalpel or razor (sharps), and place all sharps in the designated sharps container.

- Classroom decorum must be followed at all times, especially when working with potential toxins or hazards. This means no horseplay, being courteous to others, and maintaining a quiet environment.

- Pay attention to any other rules laid out by your instructor

1.2 The importance of measurements and unit conversions

Accurate measurement is fundamental to the scientific process, as it ensures the reliability of experimental results. Measurability and repeatability are key characteristics of science; precise measurements allow experiments to be replicated and verified by others, which is essential for building a robust body of scientific knowledge. Knowing how to measure correctly also facilitates clear communication among scientists. Standardized measurement techniques enable researchers to share data and compare results universally, fostering collaboration and advancing our collective understanding. Thus, mastering accurate measurement techniques is critical for the credibility, reproducibility, and advancement of scientific inquiry.

The metric system is the accepted standard due to its ability to accomplish the measurability goal of science with straightforward measurements and ease with communicating those values with others. Length is measured in **meters**, volume in **liters**, weight in **grams**, and temperature in degrees **Celsius**. We use prefixes to indicate different sizes or scales of measurements taken within each of the measurement types, and it's always the same no matter what we're measuring. For instance, kilo- is a prefix that means 1,000. When we say that something is only two kilometers away, then we know that the distance is equivalent to 2,000 meters because "kilo-" means 1,000 (see Table 1.1 and Figure 1.1). If we were to change the measurement to volume, we would also know that 2 kiloliters equals 2,000 liters because the prefix indicates so.

Table 1.1: Metric unit scale converstion chart using length (m) as an example

Length	km	m	dm	cm	mm
kilometer (km)	1	1,000	10,000	100,000	1,000,000
meter (m)	0.001	1	10	100	1,000
decimeter (dm)	0.0001	0.1	1	10	100
centimeter (cm)	0.00001	0.01	0.1	1	10
millimeter (mm)	0.000001	0.001	0.01	0.1	1

Scientists prefer the metric system because it simplifies changing between different scales of a type of measurement, known as **unit conversions**. Converting between different scales of any metric unit involves simply multiplying or dividing by 10 or its multiples, depending on the difference between the two scales (**Figure 1.1**). In contrast, conversions within the English (or Imperial) system are less straightforward. For instance, there are 36 inches in 1 yard, whereas in the metric system, there are 100 centimeters in one meter, as "centi-" denotes one hundredth. Scientists rarely use the English system to record data, but understanding how to convert between the metric and English systems can provide a better intuitive grasp of the metric units. The methods for converting between English and metric systems are also applicable within the metric system itself, making the skill of conversion highly valuable.

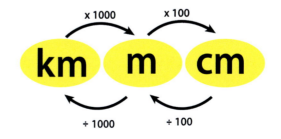

Figure 1.1: Converting between large and small scales of a unit involves changes by factors of 10.

EXERCISE: PRACTICE UNIT CONVERSIONS

In 1999, the Mars Climate Orbiter, costing $327.6 million, was lost due to a unit conversion error where one engineering team used English units and another used metric units, causing it to enter Mars' atmosphere at an incorrect altitude and disintegrate. This incident underscored the critical importance of consistent unit usage and being careful when converting units. Biologists frequently handle diverse data and often need to convert between different units. This may involve conversions within the metric system or, occasionally, between the metric and English systems. The good news is that conversions are straightforward, even if the conversion factor isn't immediately obvious. A **conversion factor** is a defined relationship between the units being converted, indicating how many of one unit equal another unit.

Example: What is the weight of a 150 lb person in kilograms? Here, we have to convert between the english and metric units.

1. Identify the conversion factor between the units. In this case, 1 kg = 2.2 lb

2. Set up a multiplication style conversion equation using the conversion factor we identified. Write the conversion factor as a fraction. This way, we can cross the units out like algebraic terms. Being able to cross the units out, as if they were variables in an algebraic equation, helps to ensure that you do the conversion correctly. This type of set-up is often called "dimensional analysis".

   The unit on the bottom cancels out with the unit you start with. If the fraction is set up incorrectly, the units will not cancel out correctly.

$$150 \, lb \times \frac{1 \, kg}{2.2 \, lb} = 68.2 \, kg$$

3. Lastly, review your answer to make sure it is logical. If you start with a larger metric unit and convert to a smaller metric unit, the numerical answer should be larger. For example, if we convert 1m into cm, the answer would be 100cm. In our example above, we know that we should expect a smaller end value because 1 kilogram is equal to more than 1 pound. If we had put 2.2 lb on top, then we would have ended up with a much larger than expected answer of 330 kg.

On the next page you will find common conversion factors needed to convert between English units and metric units (**Figure 1.2**). You will also find a diagram comparing various common temperatures between Farenheit and Celsius (**Figure 1.3**). These will be useful to you as you complete some practice problems.

Figure 1.2: Various conversion factors listed for commonly used converstions between English and metric units

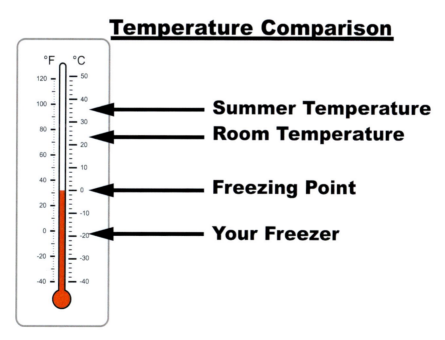

Figure 1.3: A thermometer showing a comparison between Celsius and Farenheit

The following problems serve as an important review of how to convert between different units and types of measurements. If you need help, then ask a classmate or see the instructor as soon as possible.

CONVERSION PROBLEMS:

1. Convert the following and show your work.

173 grams= _____ kg

5 ml = _____ L

14 mm = _____cm

3 lbs = _____ g

8 L = _____ml

32 gal = _____ L

56 °F = _____°C

12 km = _____ m

8 in = _____cm

2. When converting from Farenheit to Celsius, is the resulting value higher or lower?

3. What is the relationship between a meter and a kilometer?

1.3 Measuring length

The **meter** is the standard unit of length in the metric system. Length can be measured with a metric ruler or a meter stick. Using a meter stick, measure the length and width of your lab bench section and record these measurements.

Length = _____ km _____ meters _____ cm _____ mm

Width = _____ km _____ meters _____ cm _____ mm

Measuring something that is straight is fairly easy. However, there are many objects we may have to measure in the course of doing biological research that are not nice rectangular shapes. Obtain a bone from the front of the room. Measure and record the length of the bone in millimeters here:_____ Compare your measurement with that of our lab partner's. Did you make the same measurement, or was there a slight difference?

It is common for two people to measure the same item and get two slightly different numbers. This is especially the case if the object is not perfectly straight. Even when taking measurements of simple items, though, two people can be slightly different. This difference is called **measurement error**. It is important to keep in mind that there will always be slight differences from person to person or from one measurement to the next, even when something is measured by the same person. We will come back to the concept of measurement error towards the end of lab.

1.4 Measuring volume

The amount of space an object or substance occupies is referred to as volume. Volume is measured in **liters** and the most common derived unit is the milliliter (ml). Measuring volume can be much more challenging than measuring the length of an object. This is in part because the device you need to use to take a measurement needs to change more often depending on the level of accuracy needed in the measurement. Glassware, such as graduated cylinders, flasks, beakers, and pipettes can be used to measure volume (**Figure 1.4**). Each of these methods differ somewhat in how accurate they are and how much liquid they are generally used to measure. You will explore their accuracy later on in today's lab.

Flasks and beakers are best used when only an approximate measure is needed, and typically for volumes over 10ml. Graduated cylinders and pipettes provide more accuracy than flasks and beakers when measuring volume.

- **Flasks** have a smaller opening compared to beakers, making them ideal for measuring liquids that emit fumes which need to be contained or liquids that could be dangerous if splashed or spilled. They are also useful for swirling mixtures.

- **Beakers** are a wide-mouthed, cylindrical type container used for stirring, mixing, and heating liquids in laboratories.

- **Graduated cylinders** is a tall, narrow container with precise volume markings used for greater accurately when measuring liquid volumes in laboratories.

- **Funnels** help avoid spills when taking measurements and increase accuracy

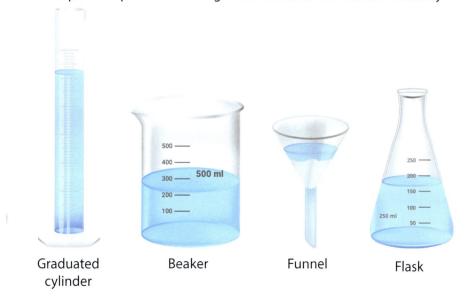

Graduated cylinder   Beaker   Funnel   Flask

Figure 1.4: Different types of instruments used to measure volume

When taking measurements of volume, read the liquid level at the meniscus. The **meniscus** is the curved surface of a liquid observed in a container, such as a graduated cylinder or beaker, caused by surface tension. In most liquids, this curve is concave, dipping down in the middle. Accurate volume measurements are taken from the bottom of the meniscus to ensure consistency and precision, as this point represents the true level of the liquid unaffected by the curvature at the edges. By reading the measurement at the meniscus, we avoid errors that could result from the liquid's interaction with the container walls, leading to more reliable and reproducible results.

To take an accurate measurement using the meniscus, begin by placing the graduated cylinder or beaker on a flat, stable surface to ensure the liquid settles evenly. Lower your eye to the level of the liquid to avoid error by ensuring your line of sight is horizontal and directly aligned with the meniscus (**Figure 1.5**). Observe the curved surface of the liquid. The meniscus is the concave curve formed by the liquid's surface tension. Read the measurement at the bottom of the meniscus. This method ensures that your measurement of the liquid volume is as accurate as possible.

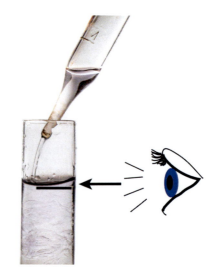

Figure 1.5: When reading the volume of liquid always look at the bottom of the meniscus

Pipettes are a special type of instrument used to measure small to moderate volumes of liquid (usually 0.5ml-50ml, see **Figure 1.6** for examples of pipettes). In general, they are a long tube that is marked on the sides with units of measure. Liquid is drawn up the tube using one of a few different ways and then dispensed where needed. There are several types of pipettes, though we will mostly use disposable or graduated pipettes.

- **Transfer or disposable pipettes** are small plastic pipettes that are used by first squeezing the bulb and then letting liquid in as you relax your grip on the bulb. It is important to squeeze the air out of the bulb before inserting it into the liquid.

- **A graduated (Mohr) pipette** is more refined than disposable pipettes and can handle larger volumes. To use it, attach the pipette to a pipette pump by holding the pipette near the top and inserting it into the pump. Use the pipette pump to draw liquid into the pipette—never use your mouth. The capacity of the pipette is printed at the top, and volume markings run from the bottom up in decreasing values. For example, on a 5 mL pipette, the first marking from the bottom might be 4. If you draw liquid to the 4 mark, you will have 1 mL of liquid. To measure 4 mL, draw up to the 1 mark.

- **Micropipettes** are a specialized form of a pipette used for measuring minute amounts of liquid. These instruments are commonly used in molecular biology for activities like protein and DNA analysis. We will cover how to use a micropipette separately in the next section because they are more complex than other types of pipettes.

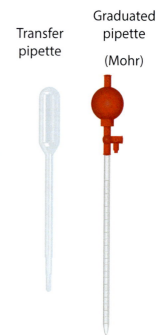

Transfer pipette

Graduated pipette

(Mohr)

Figure 1.6: Examples of pipettes

EXERCISE: USE A DISPOSABLE AND GRADUATED PIPETTE

Using a disposable pipette, transfer 10 ml of blue water into a beaker. RInse the pipette with clear water and then transfer 10 ml of yellow water into the same flask. Swirl to mix the fluids.

Repeat this process using graduated pipettes.

HOW TO USE MICROPIPETTES

A micropipette is a laboratory instrument used to measure and transfer very small volumes of liquid with high precision and accuracy (see **Figure 1.7**). Micropipettes are essential tools in many scientific fields, including biology, chemistry, and medical research. Here's a detailed description: Micropipettes can measure very small volumes in **microliters (μL)**. How small is a microliter? One microliter (μL) is 1/1,000,000 of a liter (L).

Micropipettes come in various sizes, commonly measuring 0.5-10μL (P10), 2-20μL (P20), 10-100μL (P100), 20-200μL (P200), and 100-1000μL (P1000). In today's lab, we will be using the 1000 microliter pipette, so we will focus on how to read this specific instrument.

In order for the pipette to accurately deliver the desired volume of liquid, it is important to use the instrument correctly. The same general instructions on how to use a micropipetter apply no matter what the volume of liquid is that the micropipetter can measure.

To use a micropipette properly follow the general steps outlined below:

1. Set the volume by turning the volume adjustment dial. You will find the volume on the side of the micropipette. For the P1000 microliter pipette, the top digit is the thousands, followed by the hundreds, and then the tens. The units are represented by tick-marks along the bottom of the tens band. See your instructor for exact instructions.

2. Attach a sterile plastic tip to the end of the micropipette. Tips come in boxes. To attach a tip, align the end of the micropipette with a tip in the box and gently press down until the tip is securely fitted. Once the tip is in place, avoid touching it to prevent contamination. If you need to set the micropipette down, ensure the tip does not come into contact with any surfaces or objects..

3. Aspirate the liquid sample by pushing the plunger down with your thumb to the first stop position, where you will feel resistance. Hold the plunger in this position while inserting the tip into the liquid. Slowly release the plunger, ensuring the tip remains submerged in the liquid. Once the plunger is completely released, the correct amount of liquid will be drawn into the tip of the micropipette.

4. To dispense drawn-up liquid, insert the tip of the micropipette

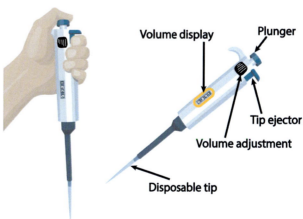

Figure 1.7: The parts of a micropipette

into the target vessel (e.g., a tube, a beaker, etc.). Push the plunger down to the first stop position. It might be necessary to push slightly past the first stop position to push all the liquid out of the tip. Withdraw the tip from the liquid before releasing the plunger.

5. To remove or change the tip, push down the eject button while the used tip is over a waste container.

Important: The plunger should only be pushed to the first stop position to take up liquid. If the plunger is pushed all the way down (beyond the resistance point to the second stop position) to take up liquid, the volume taken up will be much greater than the intended volume.

EXERCISE: USE A MICROPIPETTE

Your instructor will have you come up in small groups to have you demonstrate proper use of a micropipette. You will use one in the last part of today's lab and it is important that you use it correctly.

1.5 Measuring weight

Weight is a measure of the force exerted on an object due to gravity. Weight is a crucial characteristic in quantifying many characteristics of organisms and for conducting lab work in biology. Accurate weight measurements impact the reliability and reproducibility of molecular and organismal/ecological studies. Weight is affected by the gravitational forces which can vary based on location, unlike the closely related characteristic of mass. Because most scales do not take gravitational variation into account, it is important to realize that we are often measuring the weight of an object, rather than its mass. The reference unit for measuring weight in the metric system is the **gram (g)**. The commonly used derived units are the milligram (mg) for smaller measurements and kilogram (kg) for heaver objects.

In biology, we use a variety of scales to measure the weight of various compounds and organisms. Scales have a range of mass that they can measure accurately. In this class, many of our scales are electronic and measure up to 500 grams with an accuracy of 0.01g (**Figure 1.8**). These numbers describing the limits of a scale and its accuracy are always listed on the scale and it is important to know what they are before weighing something to avoid breaking the scale, and to avoid using the wrong scale. For instance, if you wanted to measure the weight of something that is probably around 5g, then you would not want to use a scale that is rated for 1000 kg and has an accuracy of 10g (your measurement would be worthless!).

Figure 1.8: A digital scale for measuring weight in grams

It is always important to keep the scale clean and dry, especially if it is an electronic scale. It is good to get in the habit of wiping a balance off after use and to also use a weigh boat. A **weigh boat** is a small square plastic dish that we can use to hold wet items or powdery substances, so as to avoid getting the scale dirty and can be used to transfer the material to another container. When a weigh boat is used we can take its weight into account by either weighing it first and then subtracting the weight of the boat from the final measurement, or taring the scale with the empty boat. When we tare the scale we tell the scale to automatically subtract the weight of an object from the final measurement. If the scale ever reads anything other than "0" when nothing is on the scale you can also press "tare" to re-zero the scale.

 EXERCISE: USING AN ELECTRONIC SCALE

1. Turn on the balance and remove the cover if a cover exists.
2. Select three types of beans. Make sure you have one that is large and one that is small.
3. Place the weigh boat on the scale and press "tare" to zero the scale.
4. Place each bean individually on the scale and record its weight in **Table 1.2** (on the next page).
5. Calculate the mean (average) weight of all of the beans you weighed.
6. Convert the average into milligrams.
7. If you have time, there is extra space to weigh other bean types or to get data from another group to record and compare to your own data.

Table 1.2: Weights collected from three different types of beans

Bean Type	Sample 1	Sample 2	Sample 3	Sample 4	Sample 5	Mean

QUESTIONS:

1. How much of a difference is there between the average bean weight and individual bean weight measurements?

2. Is there greater variation in bean weight if the bean is larger or smaller? How might you look at this question?

3. Can you convert the average weight of your beans from grams to miligrams? Show your work.

1.6 Exploring the accuracy of volume using weight

Percent error measures the accuracy of an experimental value by comparing it to a known or accepted value. We look at percent error to assess the accuracy and reliability of experimental results. In this exercise, you will now use your ability to weigh objects to assess which method of measuring volume is more accurate. You will measure a fixed volume (5ml) of water using each instrument listed and will weigh that measurement. From looking at measurements made across all the groups in class you will be able to calculate the percent error. A larger percent error, the less accurate the instrument is in its measure of volume. It is important to note that 1ml of water weighs exactly 1g. Therefore, your theoretical weight is 5g.

EXERCISE INSTRUCTIONS:

To investigate the accuracy of volume measurement techniques you will need to get a weigh boat, a scale, a beaker, a graduated cylinder, a graduated pipette, and a P1000 micropipette. You will also want to get a second beaker filled with some water to use. Make sure you have already received instruction on and demonstrated proper use of a micropipette before beginning.

PROCEDURE:

1. Take the weigh boat and tare the scale with it. This way, your measurement of water in the cup will be a measure of only the water's mass and not the boat.

2. Beginning with the beaker, measure out 5ml of water and add to the weigh boat.

3. Record the weight of the weigh boat with the water in **Table 1.3**.

4. Empty out the weigh boat and wipe it with a paper towel to dry it completely.

5. Now repeat steps 1-4 using the other instruments for taking volume measurements: the 10mL graduated cylinder, then a glass pipette, and finally, a micropipette. **Remember that the P1000 micropipette can only measure a maximum of 1ml. Therefore, it will be necessary for you to measure out 1ml of water 5 times to put 5ml of water in the boat.**

Table 1.3: Individual group data on the accuracy of volume measurements using different instruments

Instrument used to measure water	Water weight (g)
Beaker	
Graduated Cylinder	
Graduated Pipette	
Micropipette	

6. Record your data on the board at the front of the room.

7. Copy the class data into **Table 1.4** for each of the four methods of measuring volume of water to be weighed.

8. Using the mean for each tool from **Table 1.4** as the "experimental value", calculate the percent error (% error) usign the equation below. What is the "theoretical value"? To figure that out, consider the weight of 1 mL of water described at the beginning of this section.

$$\% \text{ Error} = \frac{\text{Experimental value} - \text{Theoretical value}}{\text{Theoretical value}} \times 100$$

9. Knowing how variable our measurements were with each device would help us decide which technique to use for a particular purpose. Calculating the range gives us a basic understanding of variability. To calculate the range, simply subtract the smallest number from the largest number. Do this for each instrument listed in **Table 1.4** and record the results in the same table.

Enter the data from your group and the other groups in your section in **Table 1.4**. For each instrument, calculate the mean, % error, and provide the range of values.

Table 1.4: Compiled class data looking at the accuracy of volume measurements made with different instruments

Instrument	Group 1	Group 2	Group 3	Group 4	Group 5	Group 6	Mean	% error	Range
Beaker									
Graduated cylinder									
Graduated pipette									
Micropipette									

Based on the data in the table above, which instrument is the most accurate for measuring 5ml? Explain your response.

Chapter 1 Post Lab Questions

1. Where are the following safety items located in the lab you are in this semester for biology 1 lab: fire blanket, first aid kit, glass disposal container, eye wash, shower.

2. Does measurement error increase or decrease with the size of the object or amount of volume being measured?

3. If I want to weigh an item on a weigh boat, but do not want the weight of the boat in my measurement what can I do?

4. Convert the following measurements:

54 kilograms = _____ g

12 inches = _____ cm

173 ml of water = _____ L

200 uL of water = _____ ml

10 cm = _____ m

1.87 mg = _____ g

17 ml of water = _____ uL

A lizard measures 178 mm long. Convert this measurement into cm. Show the units canceling out

5. You are working in the lab and begin to feel stinging sensation in your eyes. What do youdo?

6. What should you do if you are walking through lab with a test tube, drop it, and then it breaks?

Notes

Notes

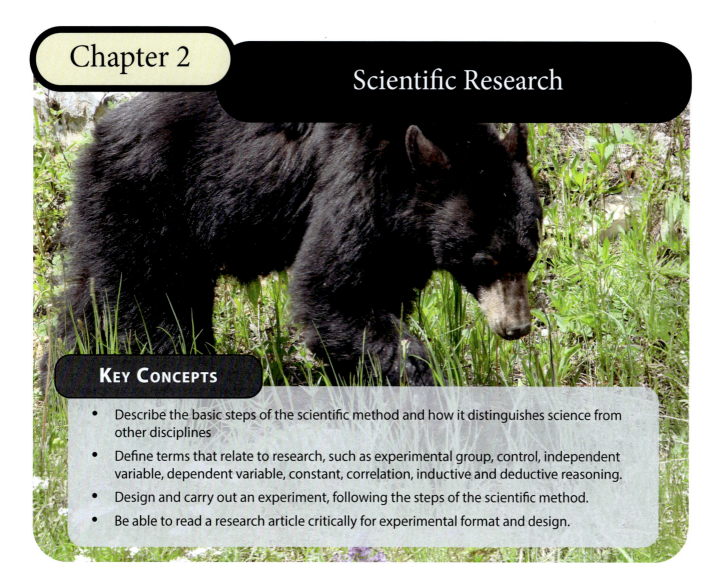

Chapter 2

Scientific Research

KEY CONCEPTS

- Describe the basic steps of the scientific method and how it distinguishes science from other disciplines
- Define terms that relate to research, such as experimental group, control, independent variable, dependent variable, constant, correlation, inductive and deductive reasoning.
- Design and carry out an experiment, following the steps of the scientific method.
- Be able to read a research article critically for experimental format and design.

Introduction

Biology is the science concerned with the study of life. However, what does it mean to be a scientific discipline? What separates science from non-scientific disciplines? The answer to what distinguishes a scientific discipline, and why biology is a science, is that biologists employ the scientific method, just like physicists, chemists, or other scientists. The **scientific method** refers to the process that is followed by any researcher in any scientific discipline to conduct research. It is a series of steps that are followed in an organized manner to find answers to questions that are asked. In 2018, some biologists used the scientific method to study whether giving people bear proof trash cans would lower conflicts with black bears, which like to forage in people's garbage. They distributed 1,110 trashcans to residents in Durango, Colorado. Results from this experiment showed the number of bear conflicts dropped by a whopping 60%!

2.1 Steps of the scientific method

STEP 1: MAKE AN OBSERVATION

The process of the scientific method begins by making observations to develop a question or series of questions. Observations can come from exploration or reading scientific papers. The scientific method is used to test the predictions of possible answers to questions about personal observations or natural phenomena. For a question to be pursued by a scientist, the observation or phenomenon must be well defined, measurable, testable, and able to be repeated by the researcher or others. Everything in the rest of the research process goes back to the question or questions posed by the researcher at the start.

STEP 2: CONSTRUCT A HYPOTHESIS

A testable hypothesis is then developed. The **hypothesis** is a proposed explanation made on the basis of limited evidence and serves as a starting point for further investigation. Based on the hypothesis, a prediction can be formulated, which describes what should be observed if the hypothesis is correct. The prediction also describes the type of data that should be gathered to test the prediction.

One of the characteristics of science that makes it different from other disciplines is that science is hypothesis driven. The goal of the scientist is to try and disprove the hypothesis. Incorrect hypotheses get discarded in favor of ones that are better. This is an example of **deductive reasoning**, where general ideas are tested and eliminated when they do not withstand scrutiny.

STEP 3: DESIGN AND CARRY OUT A STUDY

The next step is to begin the process of testing the hypothesis by performing an experiment or conducting an observational study to test the prediction. In an **experiment**, we manipulate varibles to see what might happen. In an **observational study**, we simply go gather lots of data to see what might be going on. Regardless of what type of study, we have to first take time to carefully design the study so that we make sure the data we collect will allow us to answer the question we developed together by testing the hypothesis.

One of the characteristics that both an observational study and experiment have in common is that they involve establishing or identifying different types of variables. Three basic types of variables are independent variables, dependent variables, and control variables. **Independent variables** are those factors that might explain why something is happening, which is why they are also known as explanatory variables. An independent variables is manipulated in an experiment so that it is different from one group to another group, and this allows us to see how the subject changes in response to this variation. **Dependent variables** are the factors we expect to changes in response to differences in the independent variable. For this reason, the dependent variable is commonly referred to as the response variable. For example, if the food preferences of a particular butterfly are being tested, the type of food offered will be the independent variable (what the researcher changes), and the dependent variable could be the number of times that each food is selected by an individual butterfly (the factor that is measured).

The third type of variable is the **control variable**. A control variable is a factor that should be the same for each group under study. Many times, the control variables are still measured and recorded during a study, so that the researcher can verify that they did not vary, or take any small variation into account when analyzing their data later on. Returning to our butterfly example, if the butterflies are being kept in a cage, then the temperature and time of day food is offered would be important controls. It would be a good idea to record those temperatures and times throughout the study, so that any variation in temperature or time when food was offered (in this case) could be taken into account when analyzing the data.

One of the important things to do when listing variables for a potential study is to also consider how you will measure the variables. One of the key characteristics of science is that things we study need to be measurable. So, we need to know what equipment we might need to make a measurement and what the unit of measurement will be. Some things are easy, such as taking the weight of the food we are giving to the butterflies in our example so that we give the same amount of each type of food. Counting the number of times a butterfly chooses a particular type of food, or how long the butterfly spends eating that food are also easy. However, what if we want to see if the food has an effect on the butterfly's color? These are the complexities that have to be ironed out in the design process.

When we know the types of variables we are manipulating and how we will gather the data, then we need to make sure we develop a plan for how to go about gathering the data in the same way each time. This sequence of steps to be performed or way of taking the data in an observation is called the **procedure**. Not only do we need to make sure that we take our data and make our observations in a systematic way, but we need to have enough observations to analyze at the end of the experiment. This means we need to have an idea about our **sample size**, which is the amount of data that we are collecting. Generally speaking, gathering more data in an experiment or study increases the robustness of any conclusions that are drawn. If the butterfly feeding preference study example described earlier was only completed using 10 individuals, our ability to say something about the feeding preference of that species in general would be limited. However, if the study was completed using 10,000 individuals, we could say something about the feeding preferences of the species with greater confidence.

Another aspect of experimental design closely related to sample size is **replication**, which is the number of times the experiment should be repeated in order to have confidence in the results. There should be a minimum of three replicates for statistical analysis. Continuing with the butterfly experiment as an example, we would not have much confidence in any result if we only tested the butterflies once to see what their food preferences were. However, if we repeat our experiment several times and get the same result, then we can have greater confidence in that result.

Once you are getting ready to start your experiment or study, you will need data sheets to record data in an organized manner.. Data sheets should be developed on the basis of type of data needed and with some thought about how the data will be analyzed. Although data are not tested until after the study has ended, researchers always plan how they will analyze their data before starting a study. Understanding what type of data you are collecting and how you will analyze it with statistics is critical to designing a good study.

STEP 4: ANALYZE THE DATA

Once the study is complete, the data are analyzed using statistics to determine if a difference between two groups or a trend in the data is meaningful. A meaningful trend or difference is said to be **statistically significant**. Results are summarized and displayed as graphs, tables, or as figures.

STEP 5: INTERPRET AND SHARE THE RESULTS

Once the data are analyzed, the results are interpreted. The researchers decide whether or not the results of statistical tests support the hypothesis. If the hypothesis is supported by the results, the hypothesis is accepted. If the results to not support the hypothesis, then the hypothesis is rejected.

The researchers will write a paper summarizing their interpretation of the results and talk about what other similar studies have shown in the past. This comparison to previous work is important because it helps all scientists who read the paper to understand why the results are interesting and important. The researchers will also typically present their findings at a scientific meeting. Disseminating this knowledge helps others build future research that begins where the current research leaves off to keep science moving forward.

In the process of writing up the results and sharing them with others, it is not uncommon for the original researchers to come up with new ideas. It is often the case that results of one experiment do not fully explain a phenomenon and inspire new studies. In that sense, the researchers will find themselves back at step 1, but this time making observations based on the results of their own work.

2.2 Observational studies are often starting points in biology

When a researcher makes an initial observation and formulates a question of interest, they may decide to first collect some data to see whether their initial observation might be correct, or just the product of chance. To do this, a researcher may conduct what is known as an observational study. **Observational studies** examine how an independent variable affects a dependent variable without controlling the independent variable. They differ from manipulative experiments because there are no treatment or control groups. Observational studies can be used to develop a hypothesis, which can then be refined and tested more thoroughly in an experiment where variables can be manipulated and controlled. Although observational studies are great for trying to understand what might be causing a trend or effect, it ultimately takes an experiment to determine causation.

In the following two activities, you will explore the behavior of isopods. First, you will look at data from an observational study, and then you will design an experiment to better test the hypothesis you develop from the observational study.

BACKGROUND FOR ISOPOD STUDY

Roly-polies are small, familiar "bugs" many of us played with as kids (**Figure 2.1**). Surprisingly, there are over 10,000 species of roly-polies, also known as pill bugs, which belong to the group called isopods. Isopods are part of a larger group of animals called arthropods, characterized by their hard exoskeletons. While insects, which we commonly refer to as "bugs," are part of the arthropod group, isopods are not insects but crustaceans. They are unique among crustaceans for their adaptation to terrestrial habitats, as most crustaceans, like lobsters and crabs, live in water. Their adaptation to life on land includes the development of pleopodal lungs or pseudotracheae, which allow them to breathe air. Isopods have segmented bodies that are typically flattened dorsoventrally, with seven pairs of legs that help them navigate through their habitats and avoid predators by hiding in small crevices. However, isopods exhibit a wide range of adaptations; some, like the giant isopods, can be found deep in the ocean and grow to the size of footballs.

Figure 2.1: A picture of a "dairy cow" breed of rolly polly

Terrestrial roly-polies are typically found in moist, dark environments such as under rocks, logs, and leaf litter. They earned their nickname due to their ability to roll into a ball, a defensive mechanism that protects their vulnerable undersides from predators and desiccation. This ability sets them apart from other isopods, like woodlice, which cannot roll up completely. Roly-polies primarily feed on decaying organic matter, playing a crucial role in nutrient recycling within soil ecosystems. Believe it or not, there is also a thriving collector market for various isopod species that vary tremendously in their color and patterns.

EXERCISE: ISOPOD OBSERVATIONAL STUDY

Consider the following observation:

While flipping over logs and rocks during a hiking trip, you notice that there appear to be more pillbugs present where the soil is wetter.

   a) From this observation and the background material presented to you, develop a research question you might be able to ask. You can think in terms of "does _____ affect _____?"

   b) Now take the next step and construct a hypothesis from the research question you developed. Hypotheses are predictive statements, so think about what you might expect to find when gathering data in this observational study.

   c) Identify the independent and dependent variables below that are in your hypothesis and study question.

d) With your question and hypothesis created, you decide to gather some data to see if you really do find more isopods in wetter soil. To measure soil moisture you use a small electronic sensor. Below in **Table 2.1** are the data you collected.

Table 2.1: The number of isopods found in areas that differed in moisture content of the soil

Moisture content of the soil (% water content)	Number of Isopods
4	1
6	2
7	1
16	2
20	5
23	4
21	3
37	8
38	7
46	7
10	3
30	4
32	7
33	2

e) Graph the results of your field observation study on the graph paper on the next page to analyze your results (**Figure 2.2**). The independent variable is graphed on the x-axis, and the dependent variable on the y-axis. Be sure to label the axes, including units. Also, provide a title for the graph.

What is the general trend in the data? Are there any data points that seem off from the trend?

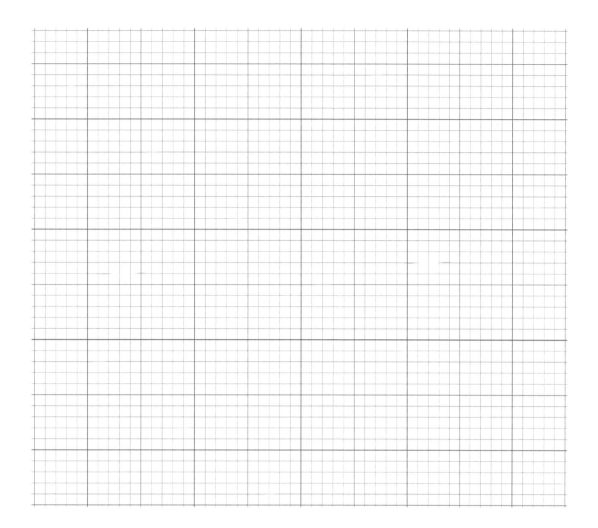

Figure 2.2: The number of isopods found under debris compared to the moisture content of the soil where they were collected

f) What is your conclusion after graphing the data?

g) Conclusions are based on a review of the data and a test of the hypothesis presented in the introduction. The outcome of one experiment often leads to more questions and additional research. Scientific experiments constantly build on one another. Based on the results from this pillbug experiment, formulate a question that could be addressed by conducting further research on this topic.

2.3 Experiments allow for more precise hypothesis testing

The observational study looking at whether more pillbugs were found in moist environments showed us that moisture could be an important factor that affects the distribution and abundance of pillbugs. However, to actually establish that the pillbugs prefer moist environments we need to conduct an experiment where we control moisture and other factors that might also influence the habitats pillbugs choose. Manipulation of variables in this manner is ultimatley what allows us to determine the cause of a phenomenon. Below are the general steps you can take to conduct a manipulative experiment to test the hypothesis concerning the preference isopods have for moist environments.

EXPERIMENTAL SET-UP QUESTIONS:

1.   What is your hypothesis?

2.   What are your dependent and independent variables?

EXERCISE: ISOPOD MOISTURE PREFERENCE EXPERIMENT

Here are the steps to follow for setting up and carrying out the isopod behavior experiment:

a)   Organize into groups of four. Wash your hands and dry them, so that you do not contaminate the isopods. Make sure to answer the questions about your experimental set up above.

b)   You will need to get the following: one behavior chamber (**Figure 2.3**), one small bowl containing 20 isopods, and one damp paper towel along with one dry paper towel.

c)   Take the damp paper towel and put it on one side of the chamber and put the dry paper towel on the other side of the chamber. Make sure the paper towels are flat against the bottom of the chambers.

d)   Place 20 isopods in the central passageway between the two chambers. Make sure the barriers are in place. Let them acclimate for two minutes.

e)   Once they have acclimated, you can remove the plastic barriers to let them choose which environment they want to head into. Record the number of in each chamber after 1 minute, 3 minutes, and 5 minutes.

f)   Repeat steps d and e two more times, recording all data in **Table 2.2**.

Figure 2.3: An illustration of the behavioral chamber you will use with the isopods to look at moisture preference

Table 2.2: Data on isopod's preference for moist or dry environments

Replicate 1			
Minutes	**Moist**	**Dry**	**Notes**
1 minute			
3 minutes			
5 minutes			

Replicate 2			
Minutes	**Moist**	**Dry**	**Notes**
1 minute			
3 minutes			
5 minutes			

Replicate 3			
Minutes	**Moist**	**Dry**	**Notes**
1 minute			
3 minutes			
5 minutes			

Average # of isopods at the end of 5 minutes	
Moist	**Dry**

g) After you have collected the data, then it is time to analyze your results. To analyze the results, we will make a bar graph using the average number of isopods that were in each side of the behavioral chamber at the end of 5 minutes (create this on the next page in **Figure 2.4**). This graph will show the similarity or difference in the average number of isopods found in the wet and dry environments.

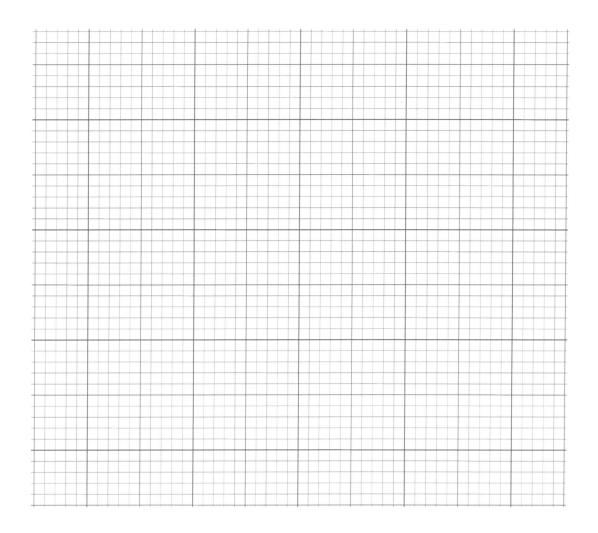

Figure 2.4: Similarity or difference in the average number of isopods found in the wet versus the dry environments

ANALYSIS QUESTIONS:

1. Was there a control in this experiment? Were there control variables?

2. Did the experiment have at least 3 replicates?

3. What are the conclusions of this experiment?

4. Did the results of this experiment support or contradict the results of the observational study?

2.4 Reading scientific research articles

A research article published in a scientific journal is an account of research reported by scientists for the purpose of sharing knowledge within the scientific community. The scientific community is both cooperative and competitive. A critical part of publishing a piece of research in a journal is having the research reviewed by other scientists. The scientific community understands that it is important to find mistakes in the thinking and experimental work of others. Therefore, research articles are reviewed anonymously by other scientists prior to being published, so that honest opinions of another researcher's work can be freely expressed. The process of reviewing research before it is published is sometimes referred to as "peer review." The journals that require peer review before publication are called peer reviewed journals. The peer review system maintains the integrity and quality of the information that appears in publication.

Scientific papers are divided into five general sections: Abstract, Introduction, Methods, Results, and Discussion. The Introduction of a scientific paper describes what is necessary for a reader to know in order to understand the importance of the work being done by the authors. It summarizes past research and describes how the current study is adding to the body of knowledge on a topic.

The Methods section describes how the authors constructed their experiments, collected specimens, analyzed data, or any other way the authors collected data to test a hypothesis. Enough detail is provided so that experiments can be repeated. Although methods sections are full of detail, some detailed steps of commonly used techniques are not described because it is assumed that other scientists will have a general understanding of the technique. There are no lists of materials or step-by-step instructions in bullet form. Everything is described in paragraph form in enough detail that another scientist could attempt to replicate that study. This section also contains descriptions of the statistical tests and computer programs used to analyze the data.

The Results section is where the authors describe the raw results from their study or experiments. They will describe averages, ranges, statistical tests, and show graphs of data. However, they do not interpret the data to describe why a value may be important, or give any context to understand the data relative to other studies (i.e. was the value lower or higher than past studies found?). These last questions that give a context to the results of statistical analyses or other findings are reserved for the discussion section.

The Discussion section is where authors will explain how their results fit into the context of previous findings. Were their values the same as other scientists found or different? Could they disprove their hypothesis, or did it turn out to be supported? They will attempt to explain why they got the results they got, and potentially come up with new hypotheses to test in future experiments as they draw conclusions from their work.

The abstract is a brief summary of the entire paper. Although the abstract is the first section in a paper, it contains relevant snippets of information from all the other sections in the paper. The goal of the abstract is to give the reader a succinct highlight of the important findings of the paper and enough background to understand the findings and questions of the research. Journals charge money for viewing their articles online. However, the abstract is always available to look at for free. In addition, scientists are busy and being able to quickly determine whether an article is pertinent to their own research is important. For these reasons, the abstract is the most important part of any paper.

Throughout every section of a scientific paper other (except the abstract) relevant papers are cited and described. The only part of a scientific paper that does not usually contain citations is the results section. These days, the number of times a paper is cited gets tracked and can potentially indicate the importance of a paper to a particular field of study. So, it is very important to properly cite other people's work!

HOW DO I SEARCH FOR SCIENTIFIC ARTICLES?

In the past, it was necessary for scientists to go to libraries and search reference cards compiled by librarians for articles written by specific authors they already knew about, or based on a few key words. Eventually, it became possible for scientists to search for journal articles across many different journals at a library of a major university that subscribed to expensive search engines (Agricola, Biosis, etc..). These days, the internet has made searching for particular scientific articles on a subject a much easier task. Although you might occasionally find it necessary to go to a library to use a specialized search engine like Agricola or Biosis, it is now possible to also use free services like Google Scholar. Scientists want their work to be found, read, and cited. So, it is also often the case that you can find a journal article on the homepage of a particular scientist (once you have found that article's name and know who the authors are).

Google Scholar can be found at https://scholar.google.com/

ASSIGNMENT:

In this assignment you will read a scientific paper. Your instructor will provide questions that you will answer about each section to get you thinking about the types of information in each.

This assignment will be due on: _____

Chapter References

Johnson, H.E., Lewis, D.L., Lischka, S.A. and Breck, S.W. (2018), Assessing ecological and social outcomes of a bear-proofing experiment. Jour. Wild. Mgmt., 82: 1102-1114. https://doi.org/10.1002/jwmg.21472

Chapter 2 Post Lab Questions

1. What is one big thing talked about in the chapter that sets science apart from other disciplines?

2. You are walking around in the rain and notice that there seem to be more frogs coming out and calling. Write a hypothesis for a study you might conduct to look at how rain might affect frogs coming out.

3. What is the main difference between an experiment and an observational study?

4. What is the dependent variable?

5. In what section of a scientific paper would you find the hypothesis?

6. In what section of a scientific paper would we comment on the importance of our findings and results?

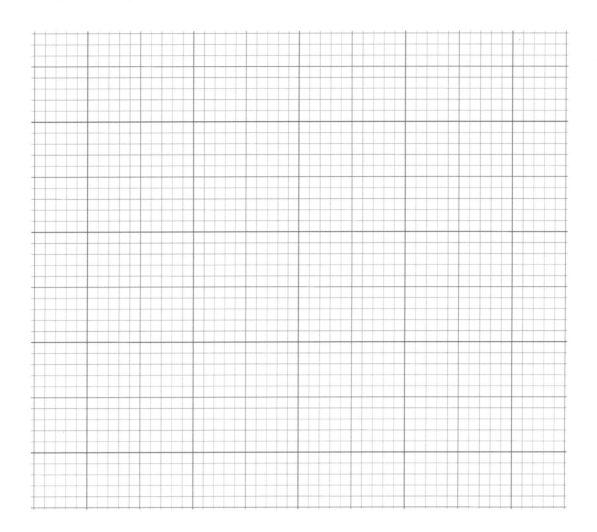

7. You are comparing the average size of lizards in three different parks (park A, B, and C). Graph the following averages in a bar graph.

--Park A: 25cm, Park B: 10cm, Park C: 33cm

--Label the X and the Y axis accordingly

Notes

Notes

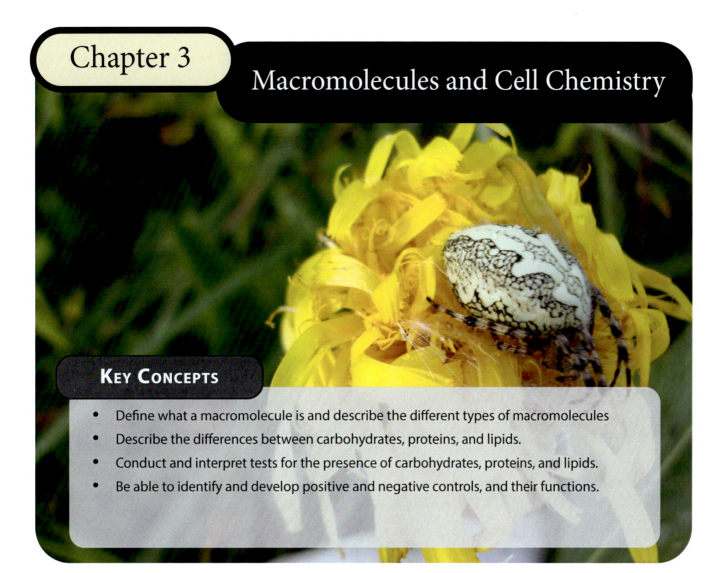

Macromolecules and Cell Chemistry

- Define what a macromolecule is and describe the different types of macromolecules
- Describe the differences between carbohydrates, proteins, and lipids.
- Conduct and interpret tests for the presence of carbohydrates, proteins, and lipids.
- Be able to identify and develop positive and negative controls, and their functions.

Introduction

Cells are fundamentally built from four primary types of biological macromolecules. These macromolecules are essential for forming cellular structures, executing cellular functions, and facilitating both intra- and intercellular communication. The four major classes of macromolecules present in living organisms are carbohydrates, proteins, lipids, and nucleic acids. Except for lipids, these macromolecules are created by linking together a few kinds of repeating units, known as monomers. The number of monomers and their identities determine the ultimate shape and function of most macromolecules. Carbohydrates are composed of monosaccharide monomers, proteins are made of amino acids, and nucleic acids are built from nucleotides. Unlike the other three groups of macromolecules, lipids are not composed of monomer units. A remarkable example of a natural extracellular macromolecule is spider silk, a protein-based fiber that exhibits incredible strength and elasticity. Spider silk surpasses even steel in tensile strength and is used by spiders to build webs, catch prey, and create shelters, showcasing the extraordinary capabilities of macromolecules in nature.

All of these macromolecules are primarily composed of carbon. Molecules are classified as organic when they consist of carbon atoms arranged in rings or long chains, with other elements such as hydrogen, oxygen, and nitrogen attached. Living organisms are largely composed of these organic molecules.

Scientists often use **chemical assays** (tests) to determine the presence of certain classes of molecules. These assays do not provide **quantitative** (numerical) results. Instead, most of these tests are **qualitative**, focusing on descriptive characteristics rather than numerical measurements. These assays are **colorimetric**, meaning they show color development when a macromolecule is detected in an unknown sample. A positive result indicates the presence of the molecule, while a negative result indicates its absence. However, qualitative colorimetric assays may not always detect the presence of a macromolecule if its concentration is too low for the assay to be sensitive enough to produce a positive result.

In this lab, assays for specific macromolecules in three organic categories—carbohydrates, proteins, and lipids—will be conducted. Each assay uses a specific chemical reagent that reacts with the target macromolecule to produce a color change. To identify the expected color change, two types of controls will be set up. The **positive control** contains the chemical to be detected and a reagent known to produce a positive result. The **negative control** contains the reagent and water but lacks the chemical detected by the reagent. These controls provide a basis for comparison with experimental samples.

For each macromolecule type tested today, you will create a positive and negative control and perform the test on an unknown. At the end of the lab, you will compile the results for all the unknowns tested by the class in **Table 3.5** on the last page.

3.1 Carbohydrates

Carbohydrates are fundamental macromolecules consisting of simple sugars or larger molecules made up of multiple sugar units. They are composed of carbon, hydrogen, and oxygen. The simplest form of carbohydrates is the **monosaccharide**, or single sugar molecule. Monosaccharides, such as glucose (the primary energy source stored in the body), fructose (fruit sugar), and ribose (a component of RNA), consist of 3 to 7 carbon atoms. These simple sugars are readily utilized by the body for energy.

When two monosaccharides are joined together, they form **disaccharides (Figure 3.1)**. Common examples of disaccharides include sucrose (table sugar) and lactose (milk sugar). When more than two monosaccharides are linked, they create **polysaccharides**. Polysaccharides are complex carbohydrates that include starch and glycogen. Starch is a crucial energy-storing carbohydrate in plants and a significant food source for humans, while glycogen is the storage form of glucose in the muscles and liver of animals.

Carbohydrate digestion occurs in the bodies of animals, where enzymes in saliva, pancreatic juice, and the intestinal lining break down both polysaccharides and disaccharides into monosaccharides. Monosaccharides such as glucose are small enough to be absorbed as nutrients and carried into the bloodstream, providing essential energy for bodily functions. This efficient breakdown and absorption process highlights the importance of carbohydrates in maintaining energy levels and overall health.

Plant cells are encased by a cell wall predominantly made of cellulose, a polysaccharide composed of glucose molecules. Although cellulose is not easily broken down by the digestive systems of animals, it serves as an important source of dietary fiber, aiding in intestinal health. Consuming adequate fiber supports digestive regularity, prevents constipation, and may reduce the risk of certain chronic diseases, such as heart disease and type 2 diabetes.

Carbohydrates play a vital role in various biological processes, not only as a primary energy source but also in cellular structure and communication. Understanding their structure and function helps elucidate their importance in both plant and animal physiology. Additionally, maintaining a balanced intake of carbohydrates is crucial for overall health, supporting energy needs, digestive health, and chronic disease prevention.

There are two primary chemical tests for detecting carbohydrates: the **Benedict's test** for small carbohydrate molecules (such as simple or reducing sugars) and the **Lugol's iodine test** for starch and larger carbohydrate molecules (more often simply referred to as the iodine test).

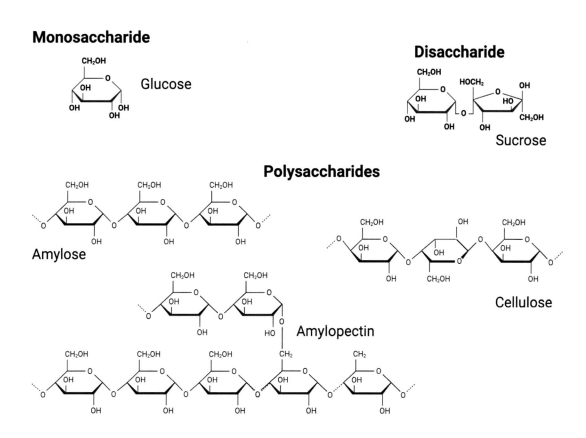

Figure 3.1: The different forms of carbohydrates (monosaccharide, disaccharide, and polysaccharide) using glucose as an example. Figure created with BioRender.com.

EXERCISE: BENEDICTS TEST FOR SIMPLE SUGARS

Benedict's test uses a copper-containing solution that reacts with all monosaccharides when heated. Disaccharides may or may not react with Benedict's solution, depending on the bonding between the monosaccharides within the compound. Polysaccharides do not test positive with Benedict's reagent unless they undergo hydrolysis, breaking them down into monosaccharides. Benedict's test not only indicates the presence of simple sugars but also provides a relative concentration. The results vary in color: a transparent blue solution indicates a negative result, green indicates trace amounts, yellow indicates a low concentration, orange indicates a moderate concentration, and red indicates a high concentration of simple sugars (**Figure 3.2**).

The Benedict's test has practical applications beyond the laboratory. In clinical diagnostics, it has been used to screen for glucose in urine samples, particularly for monitoring diabetes. In the food industry, the test helps detect reducing sugars in food products, ensuring quality control and accurate nutritional labeling.

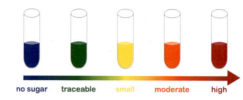

Figure 3.2: Concentration of simple sugar is indicated by different colors in the Benedict's test

PROCEDURE:

1. Label one test tube "positive", one test tube "negative", one test tube "onion", one test tube "potato", and one "unknown". Alternatively, you can label the test tubes with numbers that match against the listed contents of each test tube (i.e., tube 1 = positive, tube 2 = negative, tube 3 = onion, etc...).

2. Using a pipette, add 2 ml of each solution to be tested into individual test tubes, as indicated in **Table 3.1**.

3. Add 2 ml of Benedict's solution to each tube and mix thoroughly.

4. Record the initial color of the tube contents in **Table 3.1**.

5. Heat the labeled tubes in the boiling water bath for 15 minutes. Remove with a test tube clamp carefully, and place in your test tube rack. Record the data in **Table 3.1** and interpret your results

Table 3.1: Results from Benedict's test for simple sugars

Tube	Contents	Initial Color	Final Color	Simple Sugar Present?
Positive	Glucose			
Negative	Water			
Onion	Onion Juice			
Potato	Potato Juice			
Unknown	Unknown_____			

EXERCISE: TESTING FOR STARCH WITH IODINE

Starch, a common polysaccharide found in plants used for energy storage, can be easily detected using iodine solution (sometimes called **Lugol's iodine test**). Naturally amber or brownish in color, the iodine solution turns a dark blue-black in the presence of starch (**Figure 3.3**). This distinct color change makes the iodine test a straightforward and effective method for identifying starch in various samples. Commercially, the iodine test is used in the food industry to check for quality control and product consistency. It helps ensure that food products such as bread, cereals, and processed foods contain the appropriate levels of starch, which is essential for texture, taste, and nutritional content. For instance, in baking, the starch content can affect the dough's properties and the final product's texture. The iodine test is also used to check for adulteration in food products, ensuring that no unwanted starches are added. In agriculture the test is used to assess the starch content in crops, in the pharmaceutical industry to verify starch presence in formulations, and in the paper and textile industries to detect starch used during manufacturing.

PROCEDURE

1. Label one test tube "positive", one test tube "negative", one test tube "onion", one test tube "potato", and one "unknown".

2. Pipet 2 ml of each solution to be tested into individual test tubes as indicated in **Table 3.2**.

3. Record initial color of tube contents in **Table 3.2**.

4. Add 8 drops of Lugol's iodine to each tube, and mix thoroughly.

5. Record final color in table below. In the last column, record presence or absence of starch.

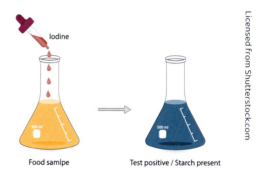

Figure 3.3: Iodine test for starch produces a dark blue to black color if positive

Table 3.2: Results from Lugol's iodine starch test

Tube	Contents	Initial Color	Final Color	Starch Present?
Positive	Starch			
Negative	Water			
Onion	Onion Juice			
Potato	Potato Juice			
Unknown	Unknown_____			

3.2 Proteins

Proteins are a diverse group of macromolecules with a variety of functions in cells, including serving as structural components, transporters, signaling molecules, and enzymes that control cellular reactions. They are the Swiss Army ™ knives of the macromolecule world. Interestingly, many venoms contain complex mixtures of proteins, each with specialized roles that contribute to the venom's potency and effects. Proteins are composed of chains of **amino acids**, each containing an NH$_2$ (amino group), a –COOH (carboxyl group), and an R (functional) group attached to a carbon backbone (see **Figure 3.4**).

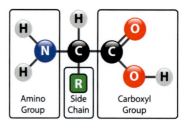

Figure 3.4: The structure of an amino acid

Despite there being only 20 standard amino acids, different sequences of these amino acids can form a vast array of proteins. The sequence of amino acids linked together by **peptide bonds** is known as the primary structure. Beyond that, secondary structure refers to localized folding of the amino acids due to hydrogen bonds forming between neighboring amino acids. Tertiary structure is the overall three-dimensional shape of a single polypeptide chain formed by additional folding and interactions among the R groups on amino acids. Lastly, most proteins are actual conglomerate molecules made up of several different chains of amino acids and this level of structure is quaternary structure.

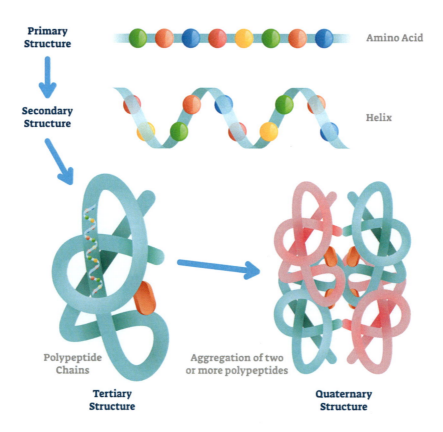

Figure 3.5: The four different levels of protein structure formed by different interactions with various parts of the amino acids

EXERCISE: TESTING FOR PROTEINS

The **Biuret test** is used to detect the presence of proteins by identifying peptide bonds. The reagent in this test, Biuret reagent, changes color based on the number of peptide bonds present. A positive result, indicating the presence of proteins, turns the solution from blue to purple (**Figure 3.6**). This shift in color is brought about by a reaction between the peptide bonds in a polypeptide chain and the Biuret reagent to form a complex. Clinically, the Biuret test is important for measuring protein levels in biological fluids such as urine and plasma, helping to diagnose and monitor conditions like kidney disease, liver disease, and proteinuria, where abnormal protein levels indicate potential health issues.

PROCEDURE:

1. Label one test tube "positive", one test tube "negative", one test tube "starch solution", and one "unknown".

2. Add 2 ml of the solutions to be tested into individual test tubes as indicated in **Table 3.3**.

3. Record initial color of contents in **Table 3.3**.

4. Add 2 ml of Biuret reagent into each tube and mix thoroughly.

5. After 2 minutes, record final color in **Table 3.3**, and in the last column, indicate whether protein was detected or not.

No protein   Proteins present

Figure 3.6: The Biuret test results and their meaning

Table 3.3: Results from the Biuret test looking for the presence of protein

Tube	Contents	Initial Color	Final Color	Protein Present?
Positive	Albumin			
Negative	Water			
Starch	Starch			
Unknown	Unknown_____			

3.3 Lipids

Lipids are a broad group of macromolecules commonly associated with oily or greasy substances. Unlike other macromolecules, *lipids are not composed of monomers that form long chains.* Instead, they vary widely in structure and function but share the common trait of being non-polar and insoluble in water, while being soluble in nonpolar solvents such as ether or chloroform. Examples of lipids include pigments like chlorophyll and hormones such as testosterone and estrogen, which play crucial roles in various biological processes. Another important class of lipids is phospholipids, which are key components of cell membranes and play a critical role in cell structure and function as the basis of cellular membranes.

Lipids that contain **fatty acid chains** (hydrocarbon chains with a carboxyl (–COOH) group at one end) combine with glycerol to form glycerides, which are important for energy storage. When three fatty acid chains attach to a glycerol backbone, they form triglycerides (**Figure 3.7**). **Triglycerides** are commonly referred to as "fats" and "oils," and their properties depend on the specific fatty acid chains attached to the glycerol molecules.

Saturated fats, such as lard, butter, or bacon fat, have fatty acids with no double bonds, containing the maximum number of hydrogen atoms and straight chains that pack closely together. In contrast, **unsaturated fats** have at least one double bond in their fatty acid chains, resulting in fewer hydrogen atoms and bent chains that cannot pack tightly toghether. This structural difference gives unsaturated fats a lower melting point compared to saturated fats.

The presence of lipids can be determined using the **Sudan III test**. When Sudan III reagent is applied to a sample containing lipids, the color changes from light pink to a darker red (**Figure 3.8**). This occurs because Sudan III is a lipid-soluble dye that is readily absorbed and concentrated by lipids, indicating their presence in the sample. In biological and medical research, the Sudan III test is used extensively to stain and identify lipids in tissues and cells, aiding in the study of lipid metabolism and related diseases.

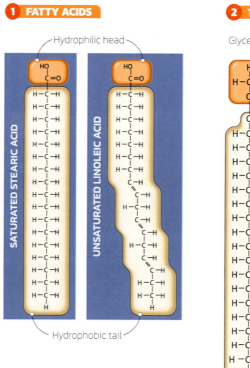

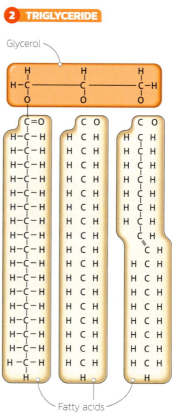

Figure 3.7: The structure of a triglyceride that is made up of different types of fatty acids joined to a glycerol backbone.

EXERCISE: LOOK FOR THE PRESENCE OF LIPIDS

In this exercise you will use Sudan III to look for the presence of lipids in the following samples and also look for it's presence in your unknown sample.

1. On a piece of filter paper, draw three circles using a pencil. Label one "positive", one "negative, and one "unknown".

2. Apply a drop of each substance to its respective circle and let it air dry in a small weighing boat.

3. Using a dropper, add a few drops of Sudan III onto the filter paper so that it covers the surface of each circle. Record initial appearance in Sudan III Initial column of **Table 3.4**.

4. Allow the paper to dry for 1 to 3 minutes, then observe the results and record in the Sudan III Final column of **Table 3.4**. Compare unknown results with the positive and negative controls. In the last column, record presence or absence of lipids.

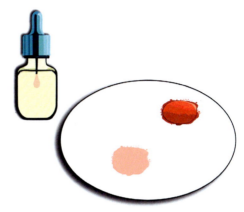

Figure 3.8: The Sudan III test being applied to samples on a piece of filter paper. The red blot is the positive result and the pink is negative

Table 3.4: Results from testing samples for the presence of lipids with Sudan III

Tube	Contents	Initial Color	Final Color	Lipids Present?
Positive	Cooking Oil			
Negative	Water			
Unknown	Unknown_____			

3.4 Identifying the unknown saples

Transfer all of your results for your unknown sample into a column in **Table 3.5** below. Indicate whether each test was positive or negative for each unknown. If a test revealed the relative amount of a substance in your sample, then also specify the concentration detected (e.g., the Benedict's test). Gather results from the rest of the class and enter them in **Table 3.5** as well. Using the positive and negative controls for each assay, determine which molecules are present in each unknown solution. Write your interpretations and your reasoning behind them on the provided lines.

Table 3.5: Results from testing the unknown samples

Tube	Unknown 1	Unknown 2	Unknown 3	Unknown 4	Unknown 5	Unknown 6
Carbs- simple						
Carbs- starch						
Proteins						
Lipids						

1. Unknown sample 1 guess: _____

   Why?_____

2. Unknown sample 2 guess: _____

   Why?_____

3. Unknown sample 3 guess: _____

   Why?_____

4. Unknown sample 4 guess: _____

   Why?_____

5. Unknown sample 5 guess: _____

   Why?_____

6. Unknown sample 6 guess: _____

   Why?_____

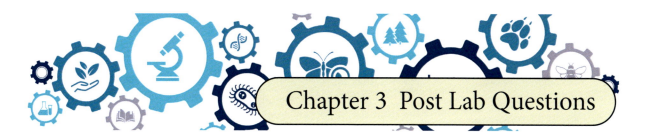

Chapter 3 Post Lab Questions

1. What is the test for simple sugars called? What is somewhat unique about what the test can tell you? How would you interpret a green test tube?

2. What does a positive result for Lugol's iodine test for starch look like?

3. What can you say about the presence of these types of carbohydrates in onions and potatoes?

4. If you add Sudan III to a small spot of liquid and see it turn red, what does it indicate?

5. What is the test for proteins called? How would you interpret a result that was light blue?

6. What was your unknown? Did you guess it correctly, or not? Was there something you missed in testing the unknown, or what was something that did not make sense?

Notes

Notes

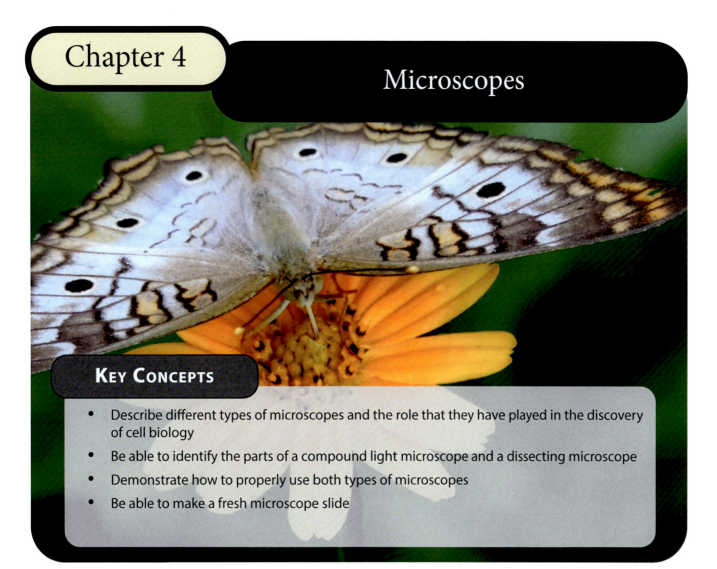

Microscopes

Introduction

All living things are composed of cells. This is a basic fact that we take for granted in the 21st century. For the longest time, the notion that organisms were made up of smaller units was unfathomable. It was ultimately the invention of the microscope that opened our eyes to cells and a whole host of microbiota (small life).

The invention of the microscope and the discovery of cells are pivotal moments in the history of science. The earliest microscopes were developed in the late 16th century by Dutch spectacle makers Hans Janssen and his son Zacharias Janssen. Their compound microscope, although rudimentary, laid the foundation for future advancements. In the 17th century, another Dutch scientist, Antonie van Leeuwenhoek, significantly improved microscope design and became the first to observe and describe single-celled organisms, which he called "animalcules." Around the same time, English scientist Robert Hooke used a microscope to examine thin slices of cork and coined the term "cells" to describe the tiny, cell-like structures he saw. These discoveries revolutionized biological science, leading to the development of cell theory and advancing our understanding of the fundamental building blocks of life.

Microscopes are now essential tools in science and come in many types, each suited for specific applications and allow us to see beyond the naked eye to varying degrees (**Figure 4.1**). They aid in studying everything from how cells work to the development of colorful butterfly wing scales.

The most common type is the **compound light microscope**, which uses visible light and lenses to magnify small objects and is widely used in biology and materials science. A **dissection microscope** is used to look at larger objects that are either barely visible to the naked eye or see small details on larger organisms during a dissection. **Electron microscopes**, such as the scanning electron microscope (SEM) and the transmission electron microscope (TEM), use electron beams to achieve much higher resolution, allowing detailed visualization of cellular structures and nanomaterials. **Fluorescence microscopes** exploit fluorescent staining to highlight specific components within a sample, making them crucial in cellular biology.

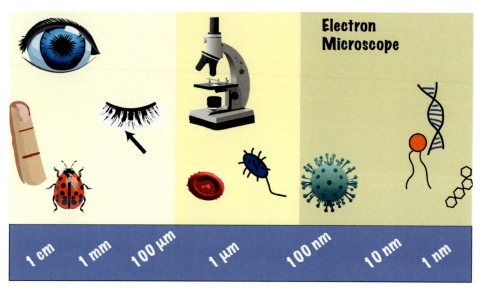

Figure 4.1: Examples off various objects that can be seen using the naked eye, a compound light microscope, and electron microscope

Once biologists started looking at cells with microscopes, they began to notice two general types of cells, based on their size and complexity. These two types of cells are prokaryotic and eukaryotic cells. **Prokaryotic cells** are smaller than eukaryotic cells and do not contain a nucleus, or other membrane-bound organelles. The DNA is found in the cytoplasm in a region called the nucleoid. Examples of life forms with prokaryotic cells include bacteria, cyanobacteria, and archaea. Unlike prokaryotic cells, **eukaryotic cells** are large and highly organized. They contain membrane-bound organelles, such as mitochondria and chloroplasts, and house DNA within a nucleus. Protists, fungi, plants, and animals are examples of eukaryotes. Although there are differences between these two cell types, it is also important to note that there are certain characteristics all cells share in common. All cells have a membrane composed of phospholipids, store genetic information in the form of DNA, and there is generally a water-based environment that fills the inside of any cell.

In addition to the two main types of cells, there are also important differences in the general appearance and structure of cells among the different groups of prokaryotes and among different eukaryotes. Most prokaryotes have distinct shapes, some of which are rod-like, circular, or spiral (**Figure 4.2**). In eukaryotes, there are distinct differences between the cells of plants and animals in shape, the presence or absence of a cell wall, or even in the types of organelles that are present inside the cell.

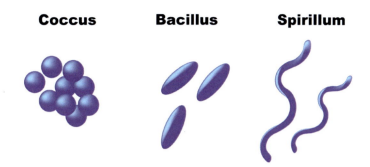

Coccus　　**Bacillus**　　**Spirillum**

Figure 4.2: Different general shapes of bacteria- sphirical (coccus), rod shaped (bacillus), and sprial shaped (spirillum)

In this lab you will be learning how to use a microscope while looking at specimens of single-celled organisms, and plant and animal tissues. Many of these specimens are already mounted on what is known as a "**prepared slide**". This type of slide typically comes from a biological supply company. The specimens have been collected, preserved, and mounted on the slide with a protective cover slip permanently glued on top of the specimen. Sometimes dyes are added to make the organism more visible, or to highlight certain features of the organism that might be of interest (i.e., a particular protein or organelle). This is why some cells will be red, blue, or some other color- they are not actually pigmented otherwise.

Prepared slides usually come with a label that describes basic information about the specimen mounted on the slide. Slide labels typically have the name of the name of the organism and a designation as to the "cut" of the preparation. Understanding how a cut was taken allows you to identify features more easily. If the organism on the slide was cut, the description of the type of cut made is important, so that a person looking at the slide understands what they are looking at. Some common designations describing how a specimen is cut are listed below and illustrated (**Figure 4.3**).

- **Whole Mount** (w.m.): The entire specimen is on the slide.
- **Cross Section** (c.s. or x.s.): Cutting a piece of a specimen at right angles (perpendicular) to the axis.
- **Longitudinal Section** (l.s.): Cutting a piece of a specimen parallel or in the same plane as the long axis of the specimen.

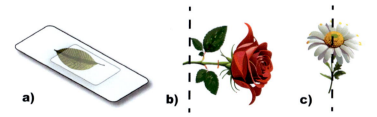

a)　　b)　　c)

Figure 4.3: The different ways of viewing organisms or parts of them on a microscope slide- (a) whole mount of a leaf (w.m.), (b) cross section of a flower stem (c.s.), and (c) a longitudinal section of a flower stem (l.s.)

4.1 Compound Light Microscopes

A) PARTS OF A COMPOUND LIGHT MICROSCOPE

A compound light microscope (abbreviated CLM) is not as difficult to use as it might first seem. One portion of the microscope holds the microscope slide in place, another is used to look at the specimen and magnify it to the degree needed to make it clearly visible, and the last portion of the microscope consists of levers and knobs that adjust the focus and amount of light to optimize the clarity of the image. It is important to know all of the parts of a microscope that ultimately work together and allow us to look into the microscopic world. This way, you can easily adjust the microscope and communicate with others about the image.

Below is a list of Compound Light Microscope parts

1. **Oculars**- The part that you look through. There is a lens within the oculars that typically magnifies the image 10x.

2. **Body Tube**- holds the oculars on the top and the nosepiece below.

3. **Arm**- is the part that connects the body tube to the base of the microscope. This is how we hold onto a microscope.

4. **Nosepiece**- Holds the objectives in place and allows you to change what objective is being used by rotating it around.

5. **Objectives**- A series of lenses that are attached to the nosepiece. There are typically four objectives. A scanning (4X), low power (10X), high power (40X), and the 100x lens, known as an oil immersion lens because it needs oil on the microscope slide to properly focus.

6. **Coarse Adjustment Knob**- This is the large knob on the side toward the base that used to make substantial adjustments in focus. **Only use the coarse adjustment knob only when the scanning objective (4x) is in place.**

7. **Fine Adjustment Knob**- Smaller knob used to make fine adjustments to the focus of the image.

8. **Condenser**- Is located just below the stage, and is used to focus the light onto the slide.

9. **Diaphragm**- An adjustable aperture that controls the amount of light coming up through the stage and onto the slide.

10. **Diaphragm Lever**- The lever that controls the diaphragm. Functionally, this lever adjusts the contrast of the image by changing the aperature.

11. **Light Source**- This is the light that shines up through the stage and to the oculars.

12. **Light adjustment dial**- Controls the intensity of the light.

13. **Base**- This is the bottom portion of the microscope that rests on the table.

14. **Stage**- This is the platform where the microscope slides are placed to examine them with the microscope.

15. **Stage Clip**- Holds onto the microscope slide.

16. **Stage Adjustment Knobs**- These knobs allow you to move the slide around smoothly on the stage. They are connected to the stage clips. One knob controls forward and backward motion and the other knob controls side to side motion.

EXERCISE: LABEL THE PARTS OF THE MICROSCOPE

Take some time to read through the parts of the compound light microscope on the previous page and label those parts on this microscope (**Figure 4.4**). You can even do this before coming to lab.

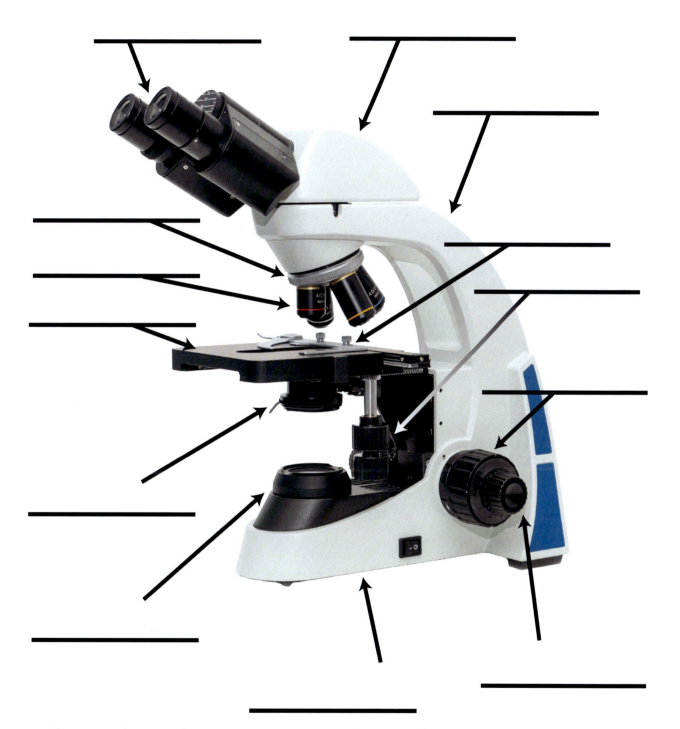

Figure 4.4: The parts of a microscope. You can complete this before class.

B) HOW TO OPERATE A COMPOUND LIGHT MICROSCOPE

Using a compound light microscope (CLM) is not difficult. To use a CLM, first make sure you have rotated the objectives so that the lowest power objective is above the stage (i.e., the shortest lens is in place to be used- usually this is the 4X lens) and lower the stage of the microscope using the coarse adjustment knob. Then, place the slide on the stage of the microscope, using the microscope stage clips to secure the slide with the label on the left side (note that the slide does not go under the stage clip). Next, with the slide roughly positioned so that the specimen is above the hole in the stage, look through the objective lens and use the coarse adjustment knobs to bring the slide into focus. Use the fine adjustment knob to fine tune the focus of the image once you have it roughly focused using the coarse adjustment knobs. If the light is too bright or too dim to see the slide clearly, then adjust the amount of light using either the diaphragm lever or the light adjustment knob. With increased magnification, more light is required. Thus, with the 4X and 10X objectives, the iris should not be fully open. When the 100X objective is used, the diaphragm should be completely open.

To increase magnification of the image, rotate the objectives around in a counter clockwise fashion using the nosepiece. The CLMs that we use are **parfocal**, which means that the image should remain relatively in focus as the magnification is increased. Therefore, you should only need to make small adjustments to the focus of an image **only using the fine adjustment knob** when increasing to higher magnifications. Always increase magnification in a stepwise manner (i.e., start at 4X, then go to 10X, to 40X, etc). That way, the image remains in focus.

To move the slide around on the stage to search the slide for the specimen of interest, use the stage adjustment knobs. The knobs enable you to move the slide smoothly in a precise way and prevent the need to touch the slide, which could negatively impact the specimen under the microscope.

EXERCISE: FOCUSING ON AN OBJECT

Follow the steps below to learn the basics of how to focus successfully in on a slide:

1. Obtain a prepared slide with the letter "e". Using the steps described above, examine the slide. Make sure to load the slide on the stage so that the "e" is right side up as you are looking at it.
2. Draw the letter "e" in the space below as it looks when viewing it through the microscope when using the 10x objective. Also explore what it looks like when you zoom in using the 40x objective to view the letter to practice increasing the magnification.
3. Describe the appearance. Is it right side up, upside down, or flipped in a different direction?

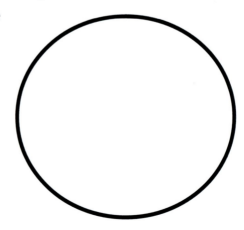

C) MAGNIFICATION

When looking through a microscope at any object, it is important to know how much the object is magnified. With two-lens systems, there is a simple method to figure the total magnification when using each objective. The magnification of each lens is written on the lens in an easily readable area. The **total magnification** is equal to the magnification of the ocular multiplied by the magnification of the objective. For each objective, determine the total magnification and write it in **Table 4.1**.

EXERCISE: CALCULATE TOTAL MAGNIFICATION

Table 4.1: Calculated total magnification values for objectives in a compound light microscope. Look at the objectives to find their magnification.

Objective	Objective Magnification	Ocular Lens	Total Magnification
Scanning			
Low Power			
High Power			
Oil Immersion			

D) UNDERSTANDING RESOLUTION

Resolution is the capacity to distinguish two points as separate entities. When using a microscope or any optical device, it becomes increasingly challenging to see closely spaced points as distinct as the image becomes more magnified. Resolution is impacted by the **aperture** of the objective lens, which is the opening of the lens and a measure of the lens's ability to take in light that helps to resolve an image. The aperture is usually indicated by a small number on the objective that can include a decimal. The larger the numerical aperture, the larger the cone of light that will be focused on the object, and higher resolution will be possible. An object can be magnified larger and still be blurry, because although magnification has increased, the resolution has not improved.

1. What is the difference between magnification and resolution?

2. What effect does increasing the magnification have on resolution (increase or decrease resolution)?

E) DEPTH OF FOCUS

Although a microscope slide might appear to be flat and the image might appear 2D, the fact is that there is a vertical dimension to the image (albeit a very small dimension). **Depth of focus** refers to the thickness of the plane of focus. In other words, the vertical distance that remains in focus at one time, which is the greatest distance through which an object can be moved while maintaining a clear image. When initially focusing on a specimen, the first portion of the specimen to come into focus is whatever is on top, followed by objects that are underneath. So, your vision moves down through a very small vertical dimension as you focus.

Below is a slide of the filamentous alga, *Spirogyra* (**Figure 4.5**). How many layers are visible? Of these layers, how many are in focus? _____

Figure 4.5: A picture of the filamentous alga *Spirogyra* as a whole mount on a microscope slide

EXERCISE: EXPLORE DEPTH OF FOCUS WITH THREADS

1. Obtain a prepared slide of several colored threads mounted together. Place the slide on the stage and focus at the point where the threads cross using the scanning objective.

2. Next, increase the magnification to the low power. Slowly focus up and down using the fine adjustment knob. How do the colors of the threads appear as you focus up and down? What color thread is:

    a. on the top?_____

    b. in the middle?_____

    c. on the bottom?_____

3. Switch to high power (40X objective). Using the fine adjustment knob, focus up and down. How does the depth of focus at high power compare with the depth of focus at low power?

F) FIELD OF VIEW AND HOW TO MEASURE OBJECTS UNDER A MICROSCOPE

The size of objects viewed under a microscope can be measured if the diameter of the viewing area and the total magnification is known. The total area that can be viewed when looking through a microscope is referred to as the **field of view** (**FOV**). As the magnification increases, the diameter of the FOV decreases. Therefore, there is an inverse mathematical relationship between the magnification and diameter of the FOV.

When viewing cells under a microscope, we can easily use the FOV to determine their average size. Simply take the diameter of the FOV and divide it by the number of cells that you count from one side to the other side across at the widest point.

For our microscopes, the FOV of the 10X objective is 1.8mm, 0.45mm for the 40X objective, and 0.18mm for the 100x objective. To measure the FOV for the 4x objective, take a plastic ruler, lay it across the stage, and use it to determine the approximate FOV for the microscope set to the 4X scanning objective. What is it in mm? _____

$$\text{Cell Size} = \frac{\text{Diameter of FOV}}{\text{Number of cells}}$$

Take a moment to practice determining the size of cells with the following two examples. The circles represent the field of view in your microscope with square and oval shaped cells. So, this first part does not require you to look through your microscope.

1. Using the low power objective (10x), calculate the length of each cell below.

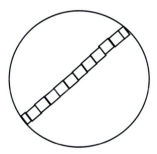

2. It isn't too hard to figure out the size of a cell when you can easily see and count how many cells fit across the field of view. However, this is often not the case. You will most likely have to use your imagination to estimate the number of cells that fit across the field of view. In this case, you are using the high power objective (40x). Apply what was described above to calculate the cell's length.

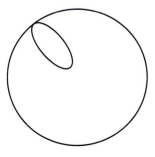

Now that you have practiced using the field of view to measure something you are looking at, it is time to put this concept into practice with real cells to determine their actual size. We will compare the size of eukaryotic and prokaryotic cells using this method.

First, let's figure out the size of eukaryotic cells by looking at an onion.

1. Obtain a prepared microscope slide of onion cells.

2. Observe the cells under 10X and 40X magnification, and decide which might be better to use if you want to determine the average cell size.

3. After you decide, count the number of cells across the diameter of the FOV lengthwise for a line of cells. Record this number below.

   Lengthwise: _____

4. Draw a quick sketch of what you see under the microscope:

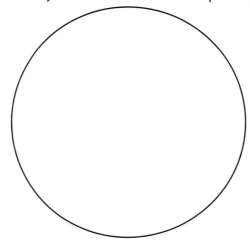

5. Calculate the average length and width of the cells you are looking at under the microscope as you did in the practice problems before and show your work in the space below.

Now, look at some bacteria cells that are set up on microscopes at the front of the room. Note that the objective being used is the 100X objective. This objective is known as an oil immersion objective because it uses oil to properly focus the microscope at such a high magnification. When light travels through a glass slide and contacts air, it will bend and distort the image. Oil can be used to decrease the refraction (or bending) of light rays, preventing contact with air, and providing a clearer image. Since this lens will only focus effectively with oil, using this lens without oil will not be useful when viewing specimens at 1000X total magnification. NEVER rotate the 40X objective into position after oil has been added to a slide. Oil will damage other lenses and distort the image observed when those lenses are used. *It is also important to use lens paper to clean the oil off the 100x objective after using it, so that you do not scratch the objective lens.*

1. How many cells do you count across the width of the FOV when looking at the bacteria under the microscope? Do you think it is possible to determine the average size of the bacteria?

2. Which cell type is bigger- the prokaryote or eukaryote?

4.2 Comparing animal and plant cells with your microscopes

Microscopes opened up the world of cellular biology. As that happened, researchers began looking at fundamentally different organisms to see how they varied at the cellular level. By magnifying specimens, these microscopes revealed similarities and differences in the types of organelles, cellular organization, and overall morphology between plant and animal cells beyond what the naked eye could ever hope to see. Plant and animal cells differ primarily in their structures and functions. **Plant cells** have a rigid cell wall, chloroplasts for photosynthesis, and large central vacuoles. **Animal cells** lack these features, but have centrioles and smaller vacuoles.

A) ANIMAL CELLS

We will explore the differences between plant and animal cells first by looking at some animal cells. One of the easiest to examine are cheek cells from the interior portion of your mouth. Fortunately, we already have some cheek cells on prepared slides for you to examine.

PROCEDURE:

1. Obtain a prepared cheek cell slide
2. Locate the cells when looking through your microscope using the scanning power (4x) objective. Once located, increase the magnification to 10x and then to 40x.
3. In the circle below, sketch what one of these cheek cells looks like. Label the nucleus, cytoplasm, and plasma/cell membrane.
4. Remember that eukaryotic cells have lots of organelles that prokaryotic cells lack completely. What organelles are you able to see?

In multi-cellular organisms, a group of cells that work together to perform a specific function is known as a **tissue**. Animal organs often consist of multiple tissue layers, with vertebrate bodies primarily composed of four tissue types: *epithelial, connective, muscle, and nervous tissue.* Each type has unique structures and functions, essential for the body's health and performance.

Examine a slide of a cross-section of the intestine to observe its tissue layers. The inner layer features red-stained, columnar-shaped **epithelial tissue** cells, which include goblet cells that secrete mucus. Compare these cells with the cheek cells we previously observed, another type of epithelial tissue. The middle layer, with blue-stained cells, represents **connective tissue**, while the red-stained, horizontal cells indicate smooth **muscle tissue**. Most cells also display a darkly stained nucleus. In the circle below, draw and label the tissue layers of the rat intestine as they appear under the microscope's 40x objective.

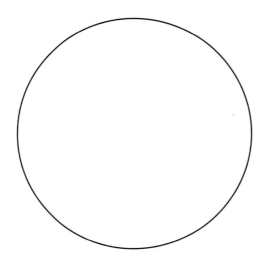

B) PREPARING FRESH SLIDES (WET MOUNTS)

There are many times in biology that we might want or need to make a fresh microscope slide from scratch to observe specimens under a microscope. This is called a **wet mount slide**. To prepare a wet mount slide, we place a drop of liquid, such as water or a staining solution, on a slide, followed by the specimen. A thin piece of plastic or glass called a **coverslip** is then carefully placed over the liquid to create a thin, even layer (**Figure 4.6**). Wet mounts are particularly useful for viewing live or freshly collected samples, as they keep the specimen hydrated and in a more natural state. This technique is commonly used in biology and medical labs to study various microorganisms, cells, and tissues, providing a clear and detailed view of their structures and behaviors.

When you have made a fresh wet mount microscope slide it is important to realize that it will eventually dry out. Water or staining solution evaporates from the edges of the coverslip. You can add more liquid to the border of the coverslip to replenish drying slides. You can also take some petroleum jelly and apply it around the edges of the coverslip to slow the rate of evaporation to prolong the life of the fresh slide. For the purposes of our lab work in this class, it will be unlikely that you need to resort to petroleum jelly, but you may find a need in future labs to add more water to the edge of the coverslip.

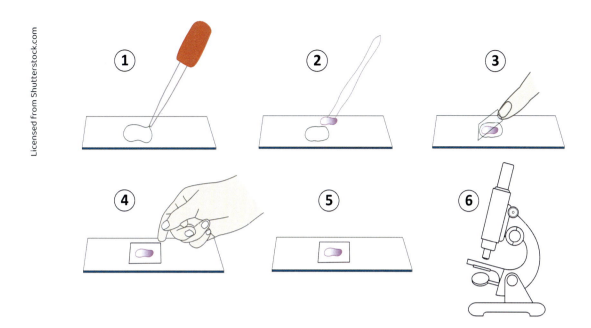

Figure 4.6: The process off making a fresh wet mount microscope slide. Add water, then add the sample, and put a cover slip at a 45 degree angle to the slide and let it fall onto the specimen.

C) LOOKING AT PLANT CELLS

It is now time to apply what we learned about how to make wet mounts to look at living plant cells. This will give you an opportunity to practice making a wet mount, which is a skill you will use many times over the course of the semester and potentially your career.

1. We will start by making a wet mount of an aquatic plant leaf. You will need to get a blank microscope slide and tear off a leaf of the plant. Put the leaf on the slide and add a couple drops of water on top. Lastly, add the coverslip.

2. Find the cells with your microscope as you have done before. Start out with the 4x objective and then increase the magnification from there.

3. In the space below, draw one of the *Elodea* leaf cells. Draw and label the nucleus, cell wall, and chloroplasts (the small green dots). You may or may not be able to see the central vacuole.

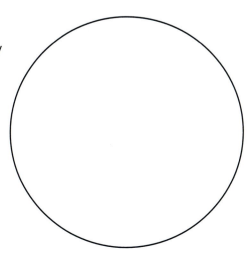

D) COMPARE PLANT AND ANIMAL CELLS

Recall that plant and animal cells have distinct structural differences. Having now examined both types, consider the following questions to compare them.

1. From your observations, what features do plant cells and animal cells share in common?

2. What features are unique to plant cells and animal cells? Which of those are you able to observe with a compound light microscope?

4.3 Using a dissecting microscope for larger specimens

Dissecting microscopes, also known as stereomicroscopes, are designed for low-magnification observation of a specimen. Unlike compound microscopes, they provide a better three-dimensional view of the specimen, making them ideal for detailed dissection and examination of larger or more complex structures. These microscopes use reflected light from above the specimen rather than transmitted light from below, allowing for the clear visualization of surface details. Specimens can be physically manipulated under magnification, since they do not have to be mounted onto a slide for observation under a dissecting microscope. Dissecting microscopes are commonly used in fields such as biology, entomology, and materials science, offering a versatile tool for both research and education. The parts of a dissecting microscope are similar to the parts of the compound light microscope, though there are fewer parts (**Figure 4.7**).

Here is a list of the different parts of a dissecting microscope. Use the list to label the microscope on the next page. You can do this before your lab session.

1. **Oculars**- The two eyepieces you look through that also help to magnify the image, typically by 10x.
2. **Binocular/microscope head**- This is the upper part of the microscope that the oculars attach to and then attaches to the rest of the lenses that magnify the image.
3. **Body tube or housing**- This holds the lenses that magnify what you are looking at along with the oculars. These lenses are analagous to the objectives on a CLM.
4. **Magnifying adjustment knob**- This is found on the side of the microscope near the oculars and allows you to change the magnification, typically only up to 35x.
5. **Arm**- provides support that attaches the body tube to the base of the microscope and allows you to hold the microscope easily.
6. **Focus adjustment knob**- Allows you to adjust the focus of the image you have in the field of view. Unlike a CLM, a dissecting microscope only has one knob.

7. **Illuminator/light source-** The light that shines down on the object on the stage to illuminate it for study. Unlike a CLM, the primary light on a dissecting microscope is above the stage, rather than below the stage.

8. **Stage-** The surface that supports and holds the specimen for viewing under the microscope.

9. **Stage clips-** Metal clips that can help hold some things in place on the stage.

10. **Base-** The lower part of the microscope that sits on the table and connects to the arm.

11. **Light switch-** Turns the light on to illuminate the specimen on the stage. Depending on the type of dissection microscope, its intensity can be adjusted and there may even be a light that can illuminate from below like what exists in a CLM.

12. **Light intensity adjustment knob-** Helps you control the intensity of the light. Some, but not all dissecting microscopes will have this.

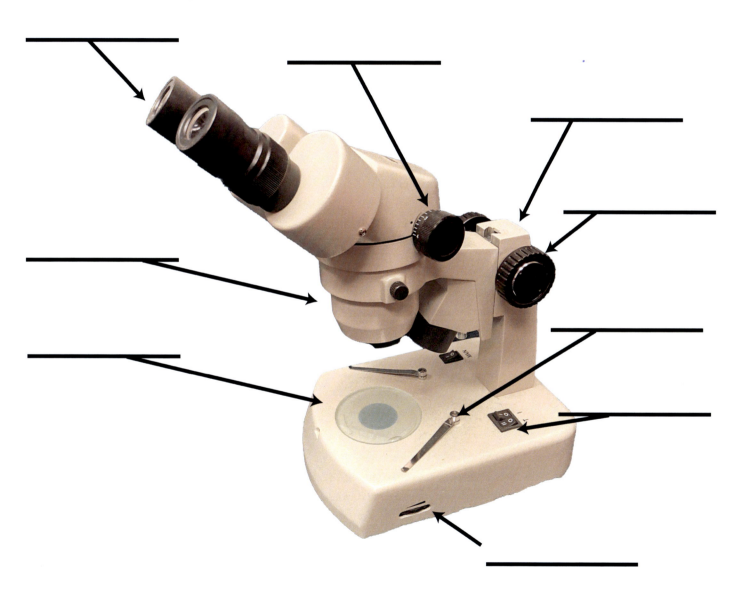

Figure 4.7: The parts of a dissecting microscope. Lines point to the parts that you should label according to the list of parts. You can complete this before class.

EXERCISE: USE A DISSECTING MICROSCOPE

Now that you have looked at the parts of a dissecting microscope and have identified those parts on a real microscope in front of you, it is time to practice using the dissecting microscope. Follow the directions below to gain some experience using it by closely examining a preserved animal specimen.

1. Obtain a plasticised specimen from the front of the room

2. Put the specimen on the stage of the microscope.

3. Turn the magnification knob so that the microscope is zoomed out all the way. You should also use the coarse adjustment knob to raise the body of the microscope up as much as possible. This is similar to how we operate a compound light microscope- we always start with the least magnification and increase from there.

4. Look through the ocular and use the coarse adjustment knob to bring the image into focus.

5. Now you can move the specimen around and adjust the magnification knob to zoom in and out to gain a better view of interesting aspects of the specimen you are examining.

6. Draw the specimen you are looking at in the space below as it appears while looking at it through the oculars.

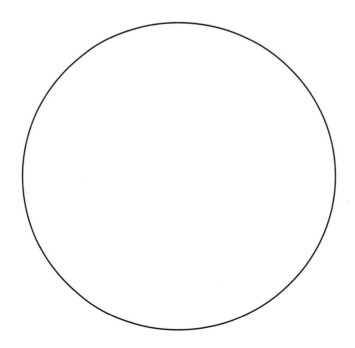

4.4 Additional practice using a compound light microscope

If you have finished the lab and have extra time, go ahead and get some additional practice using a compound light microscope.

1. Look at a prepared slide of cyanobacteria. Two types that we have available to look at are *Anabaena* and *Oscillataria*. Draw what you see.

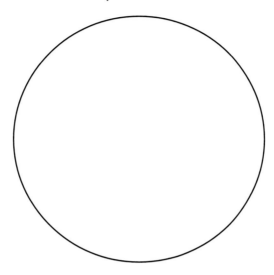

2. Make a wet mount slide of some pond water and search through the slide. Pond water teems with various microorganisms. Protozoa, such as amoebas and paramecia, can usually be found, feeding on bacteria and small algae. Additionally, algae, including single-celled phytoplankton, are present, contributing to oxygen production and forming the base of the aquatic food web. *Daphnia*, small freshwater crustaceans, are also commonly found, grazing on algae and serving as an important food source for larger aquatic animals. If pond water is NOT AVAILABLE, then search through samples of the water that the *Elodea* is sitting in.

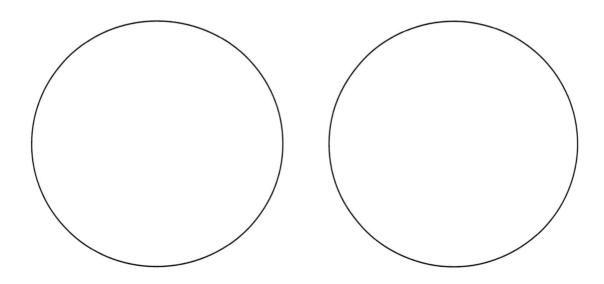

Name:_____ **Section**_____ **Date:**_____

Chapter 4 Post Lab Questions

1. Below is a picture of a plant cell. Label the structures indicated by lines. Use your textbook as a reference if needed.

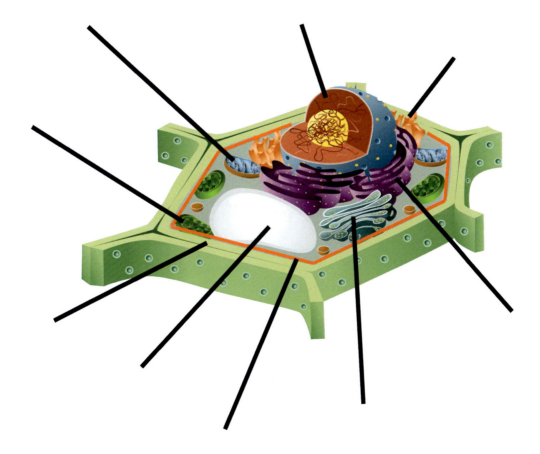

2. When comparing low power to high power, which one has the largest field of view? Explain your answer.

3. While viewing a prepared slide under low power you see a small organism at the edge of the field of view. You move to the high power lens, but the organism is no longer visible. What has occurred? How do you correct it?

4. You are looking at something under a microscope using the 40x objective. Somehow you bump the microscope and it goes out of focus. What should you do to refocus your specimen?

5. When any object is observed under the compound light microscope, which objective lens should always be used first and why?

6. You are looking at a plant cell using the 40x objective. What is the total magnification?

7. Why are some organelles not able to be seen under a compound light microscope?

8. Why were there no chloroplasts in the onion cell? Did you notice?

9. List three similarities between plant and animal cells

10. List three differences between plant and animal cells.

Notes

Notes

_____

_____

_____

_____

_____

_____

_____

_____

_____

_____

_____

_____

_____

_____

_____

_____

_____

_____

Chapter 5

Cell Membranes

KEY CONCEPTS

- Be able to describe the basics of membrane structure and function
- Understand that the rate of diffusion is affected by volume and viscosity.
- Know how concentration gradients affect the movement of water across a membrane.
- Learn the concept of tonicity and understand the effects that hypertonic, isotonic, and hypotonic solutions have on cells.

Introduction

A cell membrane, also known as the **plasma membrane**, is a crucial structure that surrounds the cell, providing a protective barrier between the intracellular environment and the external surroundings. It is primarily composed of a phospholipid bilayer, which consists of hydrophilic (water-attracting) heads and hydrophobic (water-repelling) tails, arranged in two layers. The heads orient toward the water-based environment on the outside and inside of the cell and the tails orient toward each other creating a hydrophobic interior that excludes water. This bilayer forms a **selectively permeable** membrane that regulates the movement of substances in and out of the cell (only some things can pass, not everything). Embedded within this bilayer are various proteins, including integral proteins that span the membrane and peripheral proteins that are attached to the surface . These proteins serve various functions such as transport, signal transduction, and structural support. Additionally, carbohydrates attached to proteins and lipids on the extracellular side of the membrane contribute to cell recognition and communication (**Figure 5.1**). Cholesterol molecules interspersed within the phospholipid bilayer provide stability and fluidity to the membrane, ensuring its proper function under varying conditions. The cell membrane's dynamic and complex structure is essential for maintaining cellular integrity, communication, and homeostasis.

Amphibians, like the green tree frog above, have a unique membrane structure in their skin that is essential for their survival. Their skin is not just a barrier to protect against pathogens. Unlike many animals, amphibians can take in oxygen and release carbon dioxide directly through their thin,

Chapter 5 | Cell Membranes | 73

permeable skin, which is especially important when they are in water. Their skin also has special glands that produce mucus, keeping it moist for gas exchange and providing protection against pathogens. Poison dart frogs have highly specialized poison glands that safely produce and store powerful toxins for defense. The membranes of these cells are strengthened with cholesterol and other molecules to keep toxins contained. The proteins in these membranes are specially modified to protect against damage and stay functional in a toxic environment. Additionally, the glands produce protective enzymes to neutralize the toxins' harmful effects on the cells. These adaptations show how the structure and composition of membranes can be fine-tuned to meet specific needs, highlighting the incredible versatility and adaptability of biological membranes.

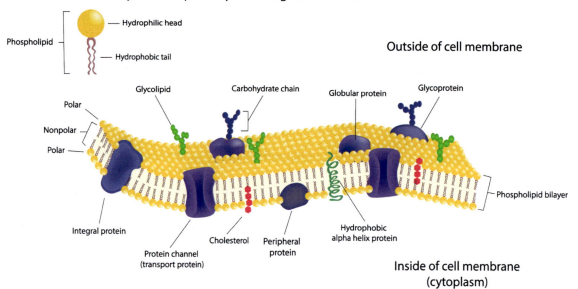

Figure 5.1: A picture of the cell membrane showing a variety of proteins embedded in a plasma membrane made up of phospholipids.

Regardless of what a specific membrane is used for, all membranes must facilitate the movement of certain substances across them. This movement can occur through passive or active transport. **Passive transport** does not require energy and relies on a concentration gradient, allowing substances to move from areas of high concentration to low concentration. Examples of passive transport include diffusion, osmosis, and facilitated diffusion, where specific proteins help move substances across the membrane (**Figure 5.2**). In contrast, **active transport** requires energy, usually in the form of ATP, to move substances against their concentration gradient, from low to high concentration. In this lab, we will be focusing on passive transport by exploring diffusion and osmosis.

Diffusion and osmosis are fundamental processes that allow movement of substances that are part of solutions across cell membranes. A **solution** is a homogeneous mixture composed of two parts: the solute and the solvent. The **solute** is the substance that is dissolved, while the **solvent** is the substance that does the dissolving, typically present in a greater amount. Both diffusion and osmosis are vital for numerous

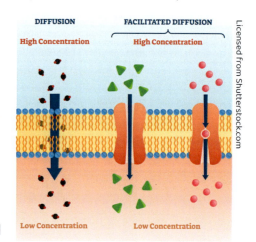

Figure 5.2: Different types of passive transport- diffusion and facilitated diffusion using proteins to get molecules across the phospholipid bilayer.

physiological processes and the maintenance of life. **Diffusion** is the movement of molecules from an area of higher concentration to an area of lower concentration until equilibrium is reached (**Figure 5.3**). This process is driven by the natural kinetic energy of molecules called **Brownian motion** and does not require additional energy. Diffusion plays a crucial role in the transport of gases like oxygen and carbon dioxide, as well as small solutes within cells and tissues.

Osmosis refers to the movement of water molecules across a selectively permeable membrane from the side with a higher water concentration to the side with a lower water concentration (**Figure 5.3**). What makes a high or low concentration of water? A solution with a high concentration of water has a lower solute concentration, whereas a solution with a low concentration of water has a higher solute concentration. So, the concentration of water is opposite the concentration of solute dissolved in the water. Osmosis is essential for maintaining cell turgor pressure, nutrient absorption, and overall cellular homeostasis.

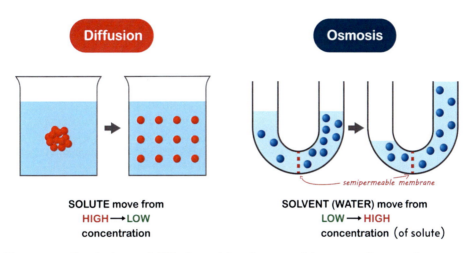

Figure 5.3: Depictions of diffusion with solute particles spreading out in a beaker of water vs. osmosis where the water moves, but not the solute particles.

The rate of diffusion down a concentration gradient is influenced by several factors. One key factor is the concentration gradient itself. The greater the difference in concentration between two areas, the faster the rate of diffusion. Temperature also plays a significant role. Higher temperatures increase the kinetic energy of molecules, thereby accelerating diffusion. The size and mass of the molecules involved affect the rate as well. Smaller and lighter molecules diffuse more quickly than larger, heavier ones. The medium through which diffusion occurs also matters, with diffusion happening faster in gases than in liquids, and even slower in solids. Additionally, the presence of a selectively permeable membrane can impact the rate, as only certain molecules may pass through, depending on the membrane's properties. Diffusion continues until equilibrium is reached. **Equilibrium** is the point where the molecules of solute are evenly distributed, and there is no net movement from one area to another. While individual molecules may still move, the overall distribution remains stable, effectively stopping the process of diffusion. Osmosis, specifically involving water movement, is affected by similar factors.

Tonicity refers to the relative concentration of solutes in two solutions separated by a selectively permeable membrane, and it determines the direction and extent of water movement between them. There are three main types of tonicity: isotonic, hypertonic, and hypotonic (**Figure 5.4**). In an **isotonic** solution, the concentration of solutes is equal on both sides of the membrane, resulting in no net movement of water and maintaining no change in cell volume. In a **hypertonic** solution, the solute concentration is higher outside the cell than inside, causing net water movement out of the cell, which can lead to cell shrinkage or crenation. Conversely, in a **hypotonic** solution, the solute concentration is lower outside the cell than inside, resulting in net water movement into the cell, which can cause the cell to swell and potentially burst, or lyse. Understanding tonicity is essential for predicting how cells will respond to their environment, especially in medical and biological contexts. In medical applications, correctly choosing isotonic, hypertonic, or hypotonic IV solutions is crucial for treating patients' hydration and electrolyte balance effectively.

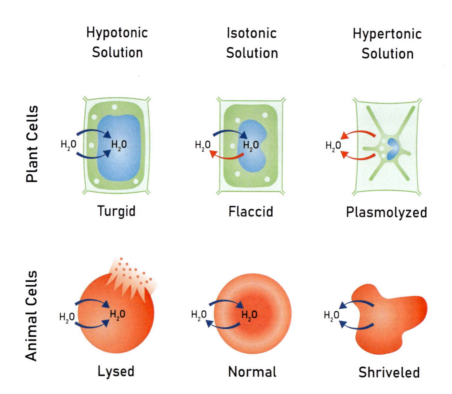

Figure 5.4: A comparison of the effects of tonicity in a typical plant cell and in a red blood cell

5.1 Exploring diffusion

Cells are typically small, largely due to the limitations on how quickly materials can be transported within them. One primary method of moving materials is diffusion, which involves the movement of molecules from an area of higher concentration to an area of lower concentration. This process relies on Brownian motion, the random movement of particles suspended in a liquid or gas, resulting from collisions with other particles. Although cells are often described as being water-based, the internal environment is more gel-like, affecting the rate of diffusion. The viscosity of this intracellular fluid can influence how quickly substances diffuse throughout the cell, impacting overall cellular function and efficiency. Because diffusion is only effective over short distances, larger cells would struggle to maintain adequate internal concentrations of essential molecules, leading to inefficient cellular function and impaired survival. Therefore, the reliance on diffusion as a primary transport method imposes a fundamental constraint on cell size.

In this exercise you will explore the process of diffusion and relate it to cells. Specifically, you will look at the differences in the rate of diffusion with respect to the volume of a container and viscosity of the medium in which something is diffusing.

EXERCISE: FOOD COLORING DIFFUSION

1. Get two small beakers and one larger beaker from the front of the room.

2. Fill one of the small beakers with corn syrup and fill the other small beaker with water. Lastly, fill the large beaker with water. You can use tap water for this exercise.

3. Predict the order that the food dye will diffuse throughout the beakers? Which will be fastest and the slowest?

4. Add 1 drop of food coloring to each beaker (use the same color in all three beakers).

5. At the times indicated in **Table 5.1**, record the extent of diffusion that has taken place in each of the beakers. After your initial observations, check back periodically, but you will want to move onto other parts of the lab while this runs.

Table 5.1: Recorded observations of how quickly food coloring diffuses.

Time	Small water beaker	Large water beaker	Small corn syrup beaker
1 min			
3 min			
60 min			

QUESTIONS:

1. In which beaker did the food coloring diffuse the fastest? Why?

2. In which beaker did the food coloring diffuse the slowest? Why do you think that was?

3. How do your observations of diffusion in the beakers relate to diffusion of substances within cells?

5.2 Osmosis and diffusion across a model cell membrane

One of the essential characteristics of a plasma membrane is its selective permeability, which allows it to regulate the passage of specific substances to maintain the intracellular environment. In this experiment, dialysis tubing will be used to simulate a cell membrane. **Dialysis tubing** functions similarly to a cell membrane by controlling which substances can cross due to its selective permeability, facilitated by tiny openings called pores. These synthetic membranes are made from regenerated cellulose fibers. Dialysis tubing is commonly used in laboratory settings to separate molecules based on size, and in medical applications such as kidney dialysis, where it mimics the function of the glomerular membrane in filtering blood. When two solutions containing dissolved particles of different molecular sizes are separated by this membrane, smaller particles can pass through the pores, while larger particles are blocked. If a molecule attempting to cross the membrane is larger than the pore size, it will be prevented from passing through.

In the following experiment you will fill dialysis bags with one of two solutions containing a mixture of starch and glucose. The concentration of glucose will be the same in each solution, but the concentration of starch will be low in one solution and high in the other. The dialysis bags will be put in an iodine solution and you will record any changes that take place to the dialysis bags throughout the experiment.

QUESTION: Based on the description above, why is this next exercise considered an experiment?

EXERCISE: MODEL CELL EXPERIMENT

1. Get two pieces of pre-cut dialysis tubing from a container of water at the front of the room. You will need to then open both ends using your index finger and thumb. The tubing dries out rapidly, so opening each end may have to be done under water.

2. Take the provided pre-cut string and tie one of the open ends. This works best when one person holds the tubing and another person ties the string.

3. With one end of the bag tied off, open the other end and place a funnel gently into the bag. Use a graduated cylinder to fill one dialysis bag with approximately 10ml of the low concentration starch solution and the other dialysis bag with 10ml of the high concentration starch solution (**Figure 5.5**). Both solutions are made with 10% glucose, so the only difference is in the starch concentration.

4. Tie the remaining open end of each bag off with another piece of string.

5. Rinse each bag off with water, gently pat them dry with a paper towel, and weigh using a balance. Record the weight of each bag on the next page in **Table 5.2** in the 0 minute row.

6. Obtain a 200ml beaker and fill it with approximately 75mL of iodine solution provided by your instructor.

7. Put the two bags in the beaker of iodine and answer the questions below.

8. After 45 minutes have elapsed, remove both bags from the beaker. Wear gloves to do this so that the iodine does not stain your skin. Rinse the bags off briefly, weigh them, and record the data in the appropriate column of **Table 5.2**.

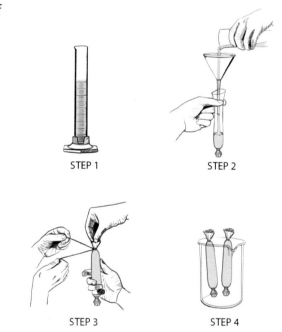

Figure 5.5: Steps for putting the model cells together from dialysis tubing.

PREDICTIONS: Think carefully about your set up. What is inside and outside of the bags? Membranes are selectively permeable, so only some things get across. Predict what each of the substances will do- cross or not cross? Explain your answer.

Table 5.2: Recorded observations from the model cell experiment

Time	Low [ ] bag weight	Bag color	High [ ] bag weight	Bag color
0 min (start)				
45 min				
Change in mass				
% change in mass				

EXPLORING AND EXPLAINING YOUR RESULTS

1. Describe the changes that occurred in each dialysis bag. Did these changes match the predictions above? If not, why might they have differed?

2. Report the direction of net water movement for each bag. Which bag had the greatest percent change in mass? Explain why.

3. Which substances did not move across the membrane? Why?

4. Which substances crossed the membrane? Why?

5.3 Exploring tonicity with living plant cells

In the previous activity, you had the opportunity to observe osmosis and diffusion in action within a model cell. Now, let's shift our focus to examining osmosis and tonicity in living cells. Changes in the solute concentration of the external environment can significantly influence whether water enters or leaves a cell. The three terms introduced earlier—hypertonic, hypotonic, and isotonic—describe these conditions. A **hypertonic** solution has a higher solute concentration outside the cell, causing more water to leave the cell and potentially leading to cell shrinkage. An **isotonic** solution has equal solute concentrations inside and outside the cell, resulting in no net water movement and maintaining cell volume. A **hypotonic** solution has a lower solute concentration outside the cell, causing more water to enter the cell, which can lead to cell swelling or bursting. When a plant cell is placed into a hypotonic solution, water moves into the cell and ultimately into the cell's central vacuole, and the cytoplasm expands. The rigid cell wall limits the amount of expansion, resulting in **turgor pressure** (pressure of the cytoplasm and cell membrane against the cell wall). A high turgor pressure will prevent further movement of water into the cell. In this activity, we will use the aquatic *Elodea* plant to observe how cells react in these different environments, providing a clear visual demonstration of osmosis and tonicity in living cells.

EXERCISE: TONICITY WITH ELODEA

In this exercise we will be treating three *Elodea* leaves with the three types of solutions (hypertonic, isotonic, and hypotonic). You will make predictions and observations for each treatment and draw a close-up of a single cell representing what you see happening to the cells in each of the three manipulations.

PREDICTIONS

1. What will happen to the cell placed in the hypertonic solution? Will there be net water movement into or out of the cell? What will that do to organelles on the inside of the cell?

2. What will happen to the cell placed in the isotonic solution? Will there be net water movement into or out of the cell? What will that do to organelles on the inside of the cell?

3. What will happen to the cell placed in the hypotonic solution? Will there be net water movement into or out of the cell? What will that do to organelles on the inside of the cell?

PROCEDURE

1. Using either forceps or your fingers, remove three green leaves from an *Elodea* sprig.

2. Make three separate wet mounts of *Elodea* leaes with clean slides (see Chapter 4, Section 4.3 for directions on making a wet mount). One wet mount should be made using the water in the Elodea container (presumably isotonic). The second wet mount should be made using distilled water from your lab bench supply basket or front of the room. The third wet mount should be made using the sodium chloride solution (NaCl) also at the front of the room.

3. Look at each leaf under the compound light microscope (CLM). You will want to pay special attention to the chloroplasts and how they are arranged inside an individual cell. Begin by looking at the leaf that is in the water from the container the *Elodea* was sitting in. This water is likely close to being isotonic. Draw an example cell in the space below and show how the chloroplasts look.

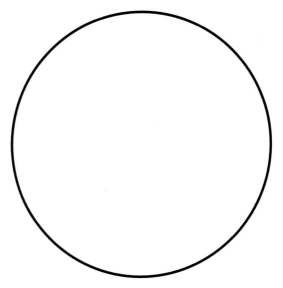

4. Now put the leaf in the hypotonic solution under the microscope. Draw what a single cell looks like in the space below. Describe how the chloroplasts are arranged. What happened to cause what you observed?

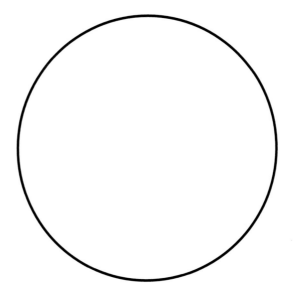

5. Lastly, put the leaf in the hypertonic solution under the microscope. Draw what a single cell looks like in the space below. Describe how the chloroplasts are arranged. What happened to cause what you observed?

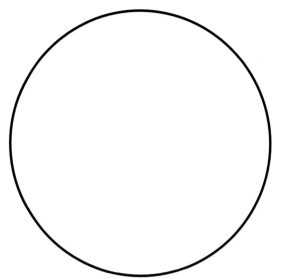

Name:_____ Section_____ Date:_____

Chapter 5 Post Lab Questions

1. How does a membrane create a selectively permeable barrier?

2. Compare and contrast diffusion and osmosis. How are they similar and different?

3. In the model cell experiment, what did the dialysis tubing represent? How did the tubing determine what passed across it?

4. How could you construct a control group for the model cell experiment?

5. If we had two chambers connected by a membrane that only lets water across and we have a hypotonic solution on one side and a hypertonic solution on the other side what would happen? In which direction would there be net water movement? Explain your reasoning.

6. Before the age of refrigeration, salting meat was essential to prevent hazardous bacteria from growing on it, which would lead to spoilage. This method was commonly used by sailors and pirates to preserve their food for long voyages at sea. Based on your experience in this lab, explain how salting food prevents bacterial growth and why it was effective for preserving meat in historical contexts like maritime journeys.

7. You were able to look at how a plant cell responds to different tonicities. How would an animal cell respond to a hypertonic or hypotonic solution? Would it be similar or different to a plant's response?

8. Examine the microscope image below showing plant cells bathed in some solution. Is this leaf sitting in an isotonic, hypotonic, or hypertonic solution? What did you base your conclusion on?

Notes

Notes

Enzymes

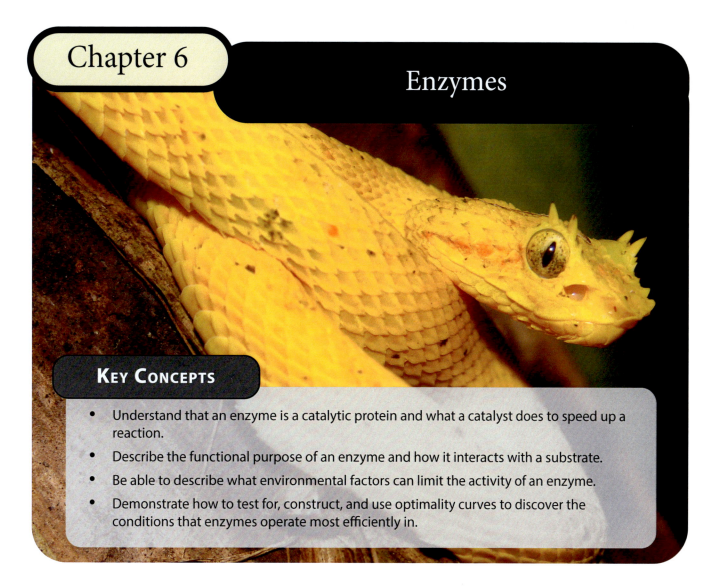

KEY CONCEPTS

- Understand that an enzyme is a catalytic protein and what a catalyst does to speed up a reaction.
- Describe the functional purpose of an enzyme and how it interacts with a substrate.
- Be able to describe what environmental factors can limit the activity of an enzyme.
- Demonstrate how to test for, construct, and use optimality curves to discover the conditions that enzymes operate most efficiently in.

Introduction

An **enzyme** is a specialized protein that significantly speeds up biochemical reactions, which are essential for the myriad of processes occurring within living cells. Enzymes speed up chemical reactions by reducing the amount of energy required to get the reaction started (**Figure 6.1**). This energy is called **activation energy**. Each enzyme is highly specific, designed to catalyze a particular reaction, ensuring the cell's chemistry is precisely regulated. This allows enzymes to play critical roles in various cellular functions, such as breaking down nutrients, synthesizing cellular components, replicating DNA, and regulating metabolic pathways. By efficiently helping to manage these biochemical reactions, enzymes maintain the delicate biochemistry that maintains life.

A fascinating example of enzymes in action is found in the venom of vipers, including the eyelash viper above. Viper venom is a complex mixture of enzymes and other proteins that work synergistically to immobilize

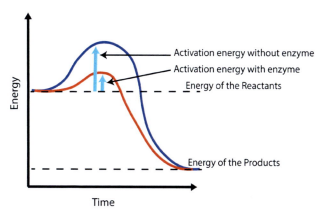

Figure 6.1: The effect that enzyme catalysts have on activation energy.

and pre-digest prey. The venom typically contains many types of enzymes such as phospholipases, which disrupt cell membranes; metalloproteases, which break down proteins and tissues; and serine proteases, which affect blood coagulation and promote tissue damage. Many types of enzymes found in viper venoms also interfere with normal cellular processes in a variety of ways. In the case of the eyelash viper, its venom contains hemotoxins and neurotoxins that cause hemorrhaging and affect the central nervous system. Needless to say, enzymes are far from being simply a component of background biochemical processes that maintain life. Regardless of what the biological role of a particular enzyme is, they all act in a similar fashion. Enzyme activity involves the formation of an **enzyme-substrate complex**, where the **substrate** (the molecule upon which the enzyme acts) binds to a specific region of the enzyme known as the active site (**Figure 6.2**). The **active site** is a unique and highly specific pocket on the enzyme's surface that fits the substrate precisely, much like a key fits into a lock. This binding induces a conformational change in the enzyme that acts on the substrate in one of many ways that speeds up the chemical reaction. Once the reaction is complete, the products are released, and the enzyme is free to bind to new substrate molecules, continuing the cycle.

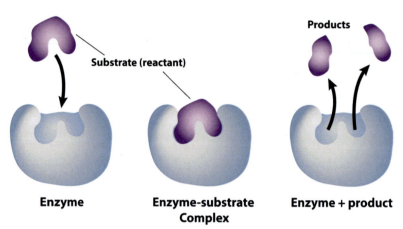

Figure 6.2: The action of an enzyme occurs in three general stages. The enzyme binds to the substrate (reactant), acts on the substrate while in the enzyme-substrate complex, and releases the modified substrate.

Enzymes are proteins, which means they have structures that are sensitive to various environmental factors such as temperature, pH, and salt concentrations, like any other protein. These factors can alter the enzyme's structure, affecting its ability to catalyze reactions efficiently. Consequently, enzymes typically operate within a narrow range of physical conditions. Each enzyme has a specific set of conditions under which it performs optimally, known as its **optimal conditions**. These conditions are influenced by the organism the enzyme is found in or the specific location of the enzyme within the organism. For instance, enzymes from bacteria in hot springs are adapted to function best at high temperatures, while enzymes in the human stomach are most effective in acidic (low pH) environments. If an enzyme is exposed to conditions outside its optimal range, it may work less efficiently and even become **denatured**, losing its functional shape. Denaturation disrupts the enzyme's quaternary, tertiary, and secondary structures, transforming it into a non-functional polypeptide chain. This loss of structure and function underscores the importance of maintaining appropriate environmental conditions for enzyme activity to ensure efficient cellular processes and overall biological function. Therefore, if we want to fully understand biological processes, then we need to learn what the optimal environments are for the various enzymes that are so critical to those processes.

One example of the importance of this can be seen in how enzyme-based medicine and supplements, such as lactase supplements, must be stored in certain conditions. The storage temperature of lactase supplements significantly affects their stability and efficacy, as high temperatures can degrade the enzyme and reduce its effectiveness. Proper storage in a cool, dry place ensures that the enzyme remains active until it is consumed, allowing it to function optimally in the body to aid digestion. When stored correctly, lactase supplements retain their activity, ensuring they provide the intended relief for lactose intolerance when taken.

We can find the optimal conditions for any enzyme by studying the rate of a reaction being catalyzed by the enzyme over a range of conditions for a particular factor (i.e., temperature, pH, etc.). Optimality curves can be constructed to understand how the physical environment an enzyme is in will affect the rate at which it catalyzes a reaction (**Figure 6.3**). To construct optimality curves, the reaction rate is plotted on the y-axis and the range of conditions for a particular factor (such as temperature or pH) is plotted on the x-axis.

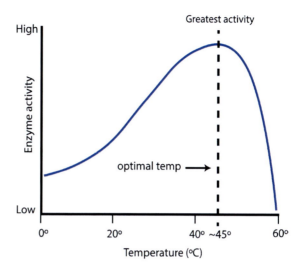

Figure 6.3: An optimality curve constructed from studying the activity of an enzyme over a range of temperatures. The optimal temperature for the enzyme is indicated by the dashed line.

In today's lab, you will be working with the enzyme catalase, which catalyzes the breakdown of hydrogen peroxide (H_2O_2) into water and oxygen (**Figure 6.4**). Hydrogen peroxide is harmful to cells because it can oxidize and denature important molecules like DNA. This is why catalase is found in the cells of nearly all organisms. For the same reasons, hydrogen peroxide is commonly used as a disinfectant and antiseptic due to its ability to kill bacteria and viruses. In high concentrations, hydrogen peroxide can be corrosive and harmful to living tissues, causing skin irritation and tissue damage. However, in low concentrations, it can be safely used for various purposes, such as wound cleaning and teeth whitening.

The breakdown of hydrogen peroxide provides an easy way to assess the rate of the chemical reaction being catalyzed by catalase because the production of gaseous oxygen causes bubbles to form. Therefore, a larger amount of bubbles indicates the chemical reaction is taking place at a greater rate of speed. The catalase for today's lab will come from potato cells, though we could use any living tissue.

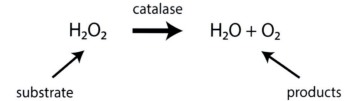

Figure 6.4: The equation showing how hydrogen peroxide is broken down by catalase.

6.1 Enzyme specificity

Here you will look at the relationship between an enzyme and its substrate. Add the contents of each test tube (which is analogous to a treatment in an experiment) described in **Table 6.1**. Record the height of the bubble column after five minutes in the table as your record of how quickly the reaction is taking place.

Create a hypothesis predicting in which tube you will see more bubbles and explain why.

Table 6.1: Data from the experiment exploring the specificity of the catalase enzyme

Test Tube	Enzyme	Substrate	Bubble column (mm)
1	2ml	Water (4ml)	
2	2ml	Sucrose (4ml)	
3	2ml	Hydrogen Peroxide (4ml)	

QUESTIONS

1. Which test tube is serving as a control? What type of control?

2. Did the enzyme react with anything other than hydrogen peroxide?

6.2 Enzyme concentration

The concentration of an enzyme can determine how quickly a reaction will occur. However, beyond a certain point, increasing enzyme concentration further has no effect on reaction speed, as the substrate becomes the limiting factor. That saturation point can vary from enzyme to enzyme. Different enzymes have varying affinities for their substrates and operate under different optimal conditions, which influences their respective saturation points. In this part of the lab you will explore how altering the amount of enzyme affects the reaction rate for the catalase enzyme. Add the contents of each test tube as described in **Table 6.2** and record the height of the bubble column in the table after five minutes.

Create a hypothesis predicting in which tube you will see more bubbles and explain why.

Table 6.2: Data from looking at how enzyme concentration affects reaction rates

Test Tube	Enzyme	Hydrogen peroxide	Water	Bubble column (mm)
1	0ml	5ml	5ml	
2	2ml	5ml	3ml	
3	4ml	5ml	1ml	
4	5ml	5ml	0ml	

1. What did you observe? What was the lowest and highest levels of output?

2. Could you discern the point at which enzyme concentration does not keep speeding up the rate of the reaction?

6.3 The effect of pH on enzyme activity

Enzymes in the human body function in a variety of pH conditions. Some are specialized to operate in highly acidic environments, such as in the stomach, and others in very basic environments. Beyond physiology, in the realm of environmental microbiology, soil-dwelling bacteria such as *Nitrosomonas*, which play a critical role in nitrogen cycling, have the enzyme ammonia monooxygenase that operates best in neutral to slightly alkaline conditions. Excess hydrogen or hydroxide ions will affect the shape of the enzyme through disrupting the hydrogen bonding within the enzyme or between the enzyme and substrate. Here you will examine the effect of pH on the activity of the enzyme. Add the contents of each test tube as shown in **Table 6.3** and record the results in the table after five minutes.

Create a hypothesis predicting in which tube you will see more bubbles and explain why.

Table 6.3: Data collected from looking at how pH affects enzyme activity

Test Tube	pH solution	Enzyme	Hydrogen peroxide	Bubble column (mm)
1	pH 3 (4ml)	2ml	7ml	
2	pH 5 (4ml)	2ml	7ml	
3	pH 7 (4ml)	2ml	7ml	
4	pH 11 (4ml)	2ml	7ml	

1. What did you observe? Which pH value had the greatest amount of enzyme activity and which had the least?

2. Construct a pH optimality curve below (label the x and y axis with factors and units where applicable):

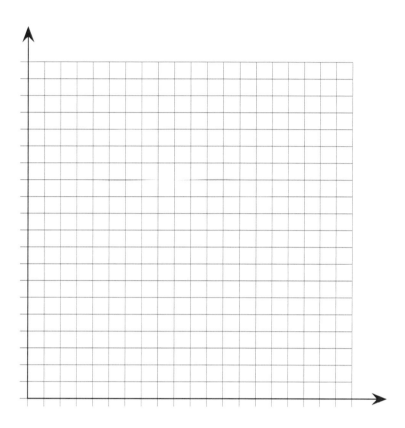

6.4 The effect of temperature on enzyme activity

Temperature varies widely in some organisms that derive their warmth from the sun (ectothermic) and for microbes that live in a number of conditions. Meanwhile, homeotherms like mammals and birds can still find themselves with body temperatures that get too hot or too cold. In all cases the metabolic processes can be affected through the effect temperature has on enzyme activity. In this part of the lab you will examine how temperature affects the rate at which the enzyme catalase catalyzes a reaction. Set up the test tubes as indicated in **Table 6.4**. Add the enzyme to the tube after letting the other contents sit in the given temperature environment for 5 minutes. Otherwise, the reaction will begin before the enzyme is affected by the temperature differences and the results will be harder to interpret. After adding the enzyme, return the tube to its given temperature environment for 5 additional minutes. Then record the bubble column height in the table below..

Create a hypothesis predicting in which tube you will see more bubbles and explain why.

Table 6.4: Data collected to look at the effect of temperature on enzyme activity

Test Tube	Temperature	Enzyme	Hydrogen peroxide	Bubble column (mm)
1	Ice bath:_____	3ml	7ml	
2	Room:_____	3ml	7ml	
3	Incubator:_____	3ml	7ml	
4	Hot:_____	3ml	7ml	

1. Construct a temperature optimality curve below. Label the x and y axis appropriately with units.

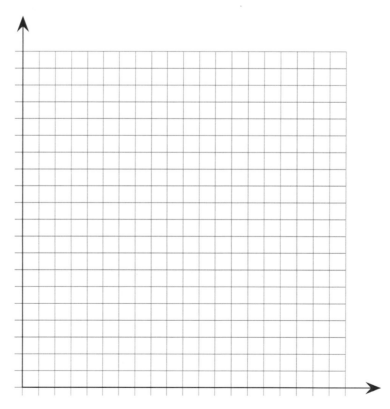

2. What did you observe with your temperature data?

6.5 The effect of substrate concentration on enzyme activity

 Now it's time to put your creativity to the test. Create your own experimental set up to test the effect of substrate concentration on enzyme activity. Write down the amounts of enzyme, hydrogen peroxide, and water added to each test tube in **Table 6.5.**

Note: All of the tubes should have the same final volume. Why is this?

Create a hypothesis predicting in which tube you will see more bubbles and explain why.

Do you need a control? If yes, then what is your control?

Table 6.5: Data collected to look at the effect of substrate concentration on enzyme activity

Test Tube	Enzyme	Hydrogen peroxide	Water	Bubble column (mm)
1				
2				
3				
4				

1. What did you observe?

2. Were there any similarities or tie-ins to looking at the enzyme concentration earlier?

Name:_____ Section_____ Date:_____

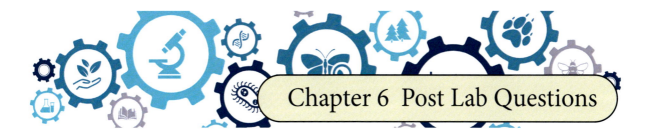

Chapter 6 Post Lab Questions

1. What is an enzyme and what does it do in a general sense?

2. What is the enzyme that we are using in today's lab? Where did it come from?

3. What is a control? What were some of the controls in this lab? Was there a control in the pH and temperature experiments? What was it?

4. What is the substrate? Where does it bind to the enzyme?

5. Compare your experiments looking at altering substrate and enzyme concentrations. Did you get similar results?

6. Having done all the work today, what can you say are the optimal conditions for catalase activity?

7. How does the structure of an enzyme relate to its function? What do we call it when the structure of a protein is lost?

Notes

Notes

Photosynthesis and Respiration

KEY CONCEPTS

- Understand the composition of light.
- Be able to describe where photosynthesis takes place.
- Explain what pigments are and how they facilitate photosynthesis.
- Compare and contrast photosynthesis and cellular respiration in plants.

Introduction

Imagine a vast, sunlit forest in the rolling hills of Appalachia, where towering oaks and stately hickories dominate the landscape, their leaves serving as miniature solar panels. These leaves capture sunlight and convert it into vital energy that fuels almost all living organisms. Beneath the canopy, animals like white-tailed deer browse through the underbrush, and timber rattlesnakes lie in wait, ready to strike. A squirrel skitters through the dappled light, foraging for acorns, unaware of the rattlesnake coiled nearby. In a swift moment, the rattlesnake strikes, and the energy stored within the squirrel is transferred to the predator. As day turns to night, the energy stored within plant cells is released, driving the ceaseless cycle of growth, repair, and renewal. In the delicate dance of life, two processes stand as the cornerstone of energy transformation in the natural world: photosynthesis and respiration. Photosynthesis allows organisms like oaks and hickories to convert solar energy into chemical energy, while respiration transforms stored chemical energy into usable energy. This chapter delves into the intricate mechanisms of photosynthesis and respiration that not only sustain the green tapestry of our planet, but also support the myriad forms of animal life, unraveling the secrets of how energy flows from the sun to every cell in our bodies.

In the web of life, every organism is either an autotroph or a heterotroph. **Autotrophs**, like the towering oaks and hickories in our Appalachian forest, produce their own food through photosynthesis, capturing sunlight and converting it into chemical energy. These plants form the foundation of the food chain. **Heterotrophs**, on the other hand, need to consume other organisms to obtain energy. The white-tailed deer, timber rattlesnake, and squirrel are all examples of heterotrophs. Deer feed on the leaves and acorns from the oaks and hickories, squirrels collect and eat these acorns, and rattlesnakes prey on squirrels for their energy. This relationship between autotrophs and heterotrophs demonstrates the flow of energy through the ecosystem and the interconnectedness of all living beings.

Regardless of whether an organism is an autotroph or heterotroph, all life forms break down chemicals to derive energy in a process called respiration. However, only autotrophs, like the oaks and hickories, have the unique ability to build carbohydrates and store chemical energy directly from the sun's energy. This process, known as **photosynthesis**, can be broadly divided into two stages: the light reactions and the dark reactions.

In the **light reactions**, which occur in the thylakoid membranes of the chloroplasts of plant cells, sunlight is captured by chlorophyll and other pigments. This energy is used to split water molecules, releasing oxygen as a byproduct and producing energy-rich molecules like ATP and NADPH. These molecules then power the second stage of photosynthesis, the dark reactions.

The **dark reactions**, also known as the Calvin cycle, do not require direct sunlight and take place in the stroma of the chloroplasts. During this stage, the energy from ATP and NADPH is used to convert carbon dioxide from the air into glucose, a type of carbohydrate. This glucose can then be used immediately by the plant for energy or stored for later use, providing a critical energy source for the plant and, ultimately, for the heterotrophs that rely on plants for food.

In this lab, we will explore both photosynthesis and respiration through exercises and experimentation. We will examine light, the pigments that are critical in converting the energy from light to chemical energy, and how the processes of photosynthesis and respiration work.

7.1 Light energy drives photosynthesis

Photosynthesis is about converting energy from the light into chemical energy. Have you ever wondered as what is light, or what causes some things to be a particular color? Our understanding of visible light is integral to various scientific disciplines, including physics, chemistry, and biology. Through the study of visible light, we have unlocked the secrets of photosynthesis, developed technologies like lasers and fiber optics, and gained insights into the nature of the universe. Like many things in science, our current understanding took hundreds of years and countless individuals studying the phenomenon to fully grasp it.

One of the early contributions to our understanding of visible light was made in the 17th century by Isaac Newton. He conducted experiments with prisms and demonstrated that white light could be separated into a spectrum of colors, each representing a different wavelength. This work laid the foundation for the field of optics. In the 19th century, James Clerk Maxwell furthered our understanding by developing the theory of electromagnetism, which described light as an electromagnetic wave. His equations unified the concepts of electricity, magnetism, and light.

Today, we know that light is part of the broader **electromagnetic spectrum** (**Figure 7.1**), which includes a wide range of electromagnetic radiation from radio waves to gamma rays. All forms of energy in the electromagnetic spectrum travel as waves, differing in wavelength. **Wavelength** is simply the distance from the crest of one wave to the crest of the next wave. The shorter the wavelength, the greater the amount of energy contained and transmitted by that wavelength. Studies of light revealed that it possesses both wave-like characteristics and particle-like properties, with photons being the fundamental particles of light.

Visible light, also known as "white light," is the segment of the electromagnetic spectrum perceptible to the human eye, spanning wavelengths from approximately 400 to 700 nanometers and encompassing all the colors we can see, from violet to red. The **color** of any object is always the function of the wavelengths of light reflected back into our eyes from the object. The complexity of white light is evident in everyday experiences, such as in the varieties of light bulbs. Some bulbs emit a "warmer" light, while others produce a "cooler" white light with less yellow. The difference is due to variations in the specific composition of wavelengths between 400 and 700nm that are emitted.

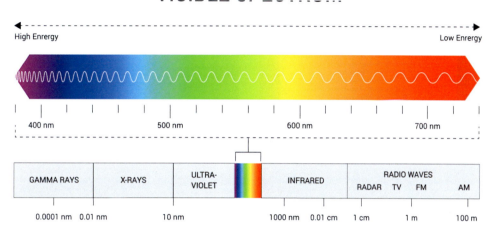

Figure 7.1: The electromagnetic spectrum consists of various kinds of energy that differ by their wavelength. Visible light is just one part of the electromagnetic spectrum. Wavelengths within the visible light spectrum can be measured in nanometers.

 ## EXERCISE: USE A SPECTROSCOPE TO EXPLORE WHITE LIGHT

Here, you will use an instrument called a spectroscope to examine the different components of white light and also see what happens when light meets objects.

1. Hold the spectroscope in front of your eyes and peer through with one eye, like you would a spyglass (**Figure 7.2**). Aim the spectroscope up at a light source. The slit of the spectroscope should be at the opposite end as your eye. You should be able to see the various colors of the rainbow along the side of the tube. Twist the eyepiece to adjust the lines of color.

2. Now have your partner hold a colored filter in front of the slit end of your spectroscope. This can be a piece of plastic or a container of colored water. Notice what happens to the colors that are along the side of the inside of the spectroscope. Describe what you observed here.

Figure 7.2: A student holding a spectroscope up toward the light

Questions:

1. What colors could you initially see when looking at a regular white light source?

2. Explain what happens when you look at the light source through the colored filter or beaker of water?

7.2 Where does photosynthesis take place?

Photosynthesis takes place in leaves, but not throughout the entire leaf. Though leaves are very thin, there is more than one cell layer that comprises a leaf (**Figure 7.3**). A leaf has an upper epidermis and lower epidermis that are both covered by a waxy cuticle. Sandwiched between are column-like palisade and round spongy mesophyll cells. Holes called stomata in the lower epidermis open and close to allow the plant to take in carbon dioxide and get rid of oxygen. Photosynthesis takes place largely in the mesophyll cells, which is where you find the chloroplasts.

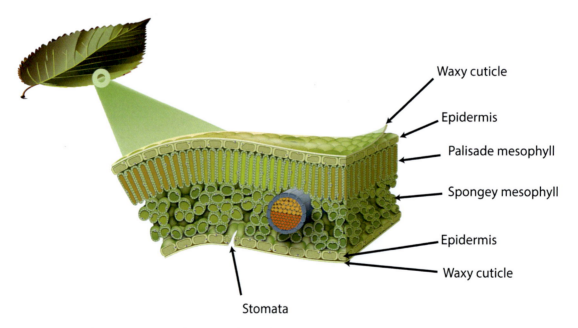

Figure 7.3: The various tissue layers of a leaf.

EXERCISE: EXAMINE A LEAF MICROSCOPE SLIDE

Find the microscope slide of the cross section of a leaf. Examine the slide under a compound light microscope. You will want to probably use the 40x objective to have a good look around. Draw what you see in the space below. Label the parts of the leaf. Can you identify all the parts in the real leaf? You can also take some time to look at a 3D model of a leaf if one is available to you.

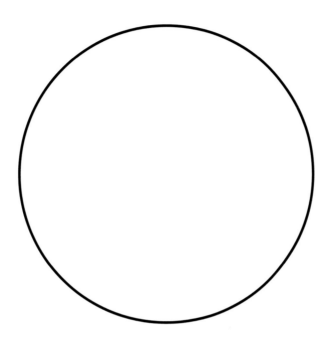

7.3 Examining pigments using paper chromatography

Light provides the energy that is converted to chemical energy in photosynthesis, but the molecules that do the converting are the pigments contained in a leaf. **Pigments** selectively absorb and reflect colors of light. Pigments are crucial components in the process of photosynthesis, acting as transducers by capturing light photons and initiating the photochemical reactions needed for energy conversion. They are involved specifically in the light reactions. The primary pigment involved is chlorophyll, which is responsible for the green color of plants. Chlorophyll absorbs light most efficiently in the blue and red wavelengths but reflects green light, giving plants their characteristic color. Additionally, accessory pigments also play important roles in addition to absorbing light from other wavelengths not covered by chlorophyll a. Common accessory pigments include carotenoids, xanthophylls, chlorophyll b, phycoerythrins, and phycobilins, (**Figure 7.4**). The carotenoid pigments such as β-carotene and xanthophyll can also protect plants by absorbing excess energy and acting as antioxidants. Anthocyanins give plants like *Coleus* their reddish color and may serve to help pollinators identify certain plants. These pigments work together to maximize the amount of light energy captured for the production of glucose and other vital organic compounds.

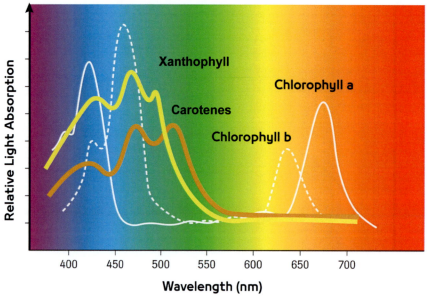

Figure 7.4: The absorption spectrum for various pigments involved in photosynthesis. It shows what wavelengths of light pigments are absorbing the most energy at.

In this exercise, we will use paper chromatography to separate and identify the different pigments found in plant leaves. **Paper chromatography** is a technique that allows us to separate complex mixtures into their individual components based on their solubility, and interactions with the solvent and the paper. By applying a sample of plant extract onto a piece of chromatography paper and using a **non-polar solvent like petroleum ether**, we can observe how the various pigments move at different rates along the paper (**Figure 7.5**). This separation occurs because different pigments have varying affinities for the solvent and the paper, causing them to travel distinct distances.

In this exercise, after separating the plant leaf pigments using paper chromatography and a non-polar solvent like petroleum ether, we will measure the distance each pigment travels from the original spot where the sample was applied. By comparing these distances to the distance traveled by the solvent front, we can calculate the Rf values for each pigment. **The Rf value**, or retention factor, is a numerical representation of how far a pigment travels on the chromatography paper relative to the solvent front. These values are useful for identifying and comparing pigments because each pigment has a characteristic Rf value under specific conditions, allowing us to distinguish between different components in the plant extract. To calculate the Rf value, simply divide the distance the pigment travels by the distance the solvent travels.

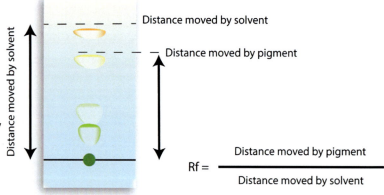

$$Rf = \frac{\text{Distance moved by pigment}}{\text{Distance moved by solvent}}$$

Figure 7.5: The end product of paper chromatography showing pigments spread out across the paper and how to calculate an Rf value

EXERCISE: PAPER CHROMATOGRAPHY

In the exercise below, you will carry out a paper chromatography exercise where you will separate pigments and then calculate the Rf values for those pigments. You will be using leaf extract that was prepared ahead of time and the solvent is petroleum ether.

PRELIMINARY QUESTION: The ability of a pigment to be dissolved in a solvent determines how quickly/far it will travel in paper chromatography. Here is some detail on the chemical structure of some pigments you may see today. Chlorophyll b has 6 polar groups, Chlorophyll a has 5 polar groups, Xanthophylls have 2 polar groups, and Carotenes have 0 polar groups. The number of polar groups determines the solubility. The more polar groups, the less soluble something will be in non-polar solvent like petroleum ether. Arrange these listed pigments in the order you expect them to be on your chromatography paper from top to bottom (least polar to most polar).

PROCEDURE

1. Obtain a piece of chromatography paper and use a pencil to draw a line across the paper about 2cm above the actual bottom. This is where we will be adding the plant extract.

2. Get the plant extract from the front of the room and use the capillary tube to begin adding the plant extract to the paper. Make three spots of extract across the line you drew with your capillary tube. Use a hair dryer to quickly dry the extract and then make another set of spots with droplets of extract on top of the three you placed before. We are adding enough extract to the paper to get a good sample of pigments. Repeat this until you have added five drops of extract to each of the three spots, drying the drops as you go.

3. Take your chromatography paper to the fume hood where jars labeled "solvent" are located. Each jar contains a mixture of petroleum ether and acetone in a 9:1 ratio. You may want to label the jar that you use. You or your instructor will add your paper to the jar under to hood and then close the lid.

4. Remove the paper from the jar once the solvent is within 1/2 cm of the top of the jar. Make note of where the solvent front is located and let the solvent dry in the fume hood. Mark where the solvent front is located with a pencil while wearing gloves and let the paper dry under the fume hood.

5. Measure how far each of the pigments traveled in mm, and record these data, along with each pigment color, in **Table 7.1**. Also measure how far the solvent traveled, and use this to calculate the Rf for each pigment. Make all your measurements for calculations in mm. Based on the colors of the pigments, and how far they traveled, try to identify each of the pigments. Use the information provided in the preliminary question above to help you. **All Rf values should be below 1.0**. What pigment was the most and least polar?

Table 7.1: Pigment information and Rf values from the paper chromatography exercise

Band color	Distance traveled (mm)	Rf Value	Pigment name

7.4 Photosynthesis and Respiration experiment

Photosynthesis and respiration are two fundamental processes that sustain life on Earth, operating in complementary cycles (**Figure 7.6**). Photosynthesis is the process by which autotrophic organisms, like plants, convert light energy into chemical energy stored in glucose. This process occurs in the chloroplasts of plant cells, utilizing sunlight, carbon dioxide, and water to produce glucose and oxygen. The plants, and other photosynthetic organisms, are able to take carbon in a biologically unusable form and convert it into a usable form- a process known as **carbon fixation**. In contrast, respiration is the process by which all living organisms, including plants and animals, break down glucose to release energy for cellular functions, producing carbon dioxide and water as byproducts. While photosynthesis captures and stores energy, respiration releases this stored energy for use by the organism. By studying these processes in *Elodea* plants here using phenol red as a pH indicator and volumeters, we can observe the dynamic interplay between photosynthesis and respiration, highlighting their critical roles in the energy balance of life.

In this experiment we need to provide plants with a source of carbon dioxide because plants require carbon dioxide for photosynthesis to take place. To add carbon dioxide we will blow bubbles into the water using a long straw. As the concentration of carbon dioxide increases in a solution, the

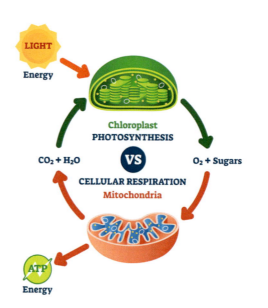

Figure 7.6: The processes of photosynthesis and respiration are complimentary. Photosynthesis builds carbohydrates and respiration breaks them down. Both processes occur in plants.

concentration of hydrogen ions ultimately increases, which decreases the pH (i.e., it becomes more acidic). This happens because CO_2 combines with H_2O to form carbonic acid, which dissociates into bicarbonate and hydrogen ions (see equation below). We will be able to see the effects of adding additional carbon dioxide to the water by using a pH indicator.

$$CO_2 + H_2O \longleftrightarrow H_2CO_3 \longleftrightarrow HCO_3^- + H^+$$

Carbon dioxide + water   Carbonic acid   Bicarbonate + hydrogen ion

Phenol red is a pH indicator and will be useful in examining photosynthesis and cellular respiration in this experiment. Phenol red is yellow under acidic conditions, red at neutral pH and pink under basic conditions. Therefore, a yellow color indicates a solution with a greater amount of carbon dioxide in our experiment, while a reddish color indicates less carbon dioxide in solution.

EXERCISE: PHOTOSYNTHESIS EXPERIMENT

Below are the instructions for setting up the experiment to look at photosynthesis and respiration in the *Elodea* plant. You will set up the experiment with either a white light bulb or green light bulb. One of the other groups close to you will set up with the other lightbulb. You will have data tables to record data from both experiments to compare and contrast the results at the end.

HYPOTHESIS: Having read through the description and methods, create a hypothesis and predict what the outcome will be in each of the three test tubes involved in this experiment.

PROCEDURE:

1. Get a large beaker and fill it with water to serve as a heat sink. This way, any heat produced by the light bulb does not affect the experiment. Any change can be solely attributed to the light energy.

2. Set your lamp up to face the beaker of water (**Figure 7.7**). Your lamp will either have a white or green light bulb. We will be comparing the results between the two light color treatments.

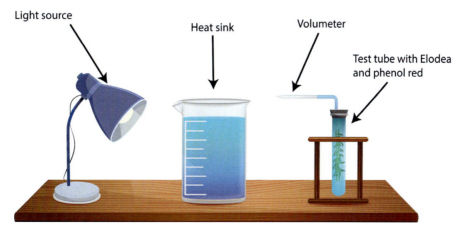

Figure 7.7: The photosynthesis experiment set up with the light source at one end and the test tubes with volumeters at the other end.

3. Get three test tubes and place a long sprig of *Elodea* in two of the test tubes. Wrap one of the two tubes containing *Elodea* in aluminium foil to block light from reaching the plant.

4. Next, ask your instructor for approximately 100 ml of dilute phenol red solution. Place the plastic graduated pipette tip into the solution and exhale carefully, blowing bubbles through the solution. Continue until the pink solution turns pale yellow. **Be careful not to inhale or swallow the solution because phenol red is toxic if consumed. Avoid coming in contact with the phenol red as well and wash immediately if any gets on your skin.**

5. Add the pale yellow phenol red solution to the three test tubes. One test tube will only contain the yellow phenol red solution. Insert the stopper of the volumeter into the test tube, leaving approximately 1-2cm of space between the liquid and the bottom of the stopper. The phenol red solution should go up and into the volumeter about 1/3-1/2 the length of the tube when the stopper is in place (see volumeter in **Figure 7.7**). Mark the starting point of the liquid on the volumeter with a wax pencil. You will measure the distance the solution moves forward or backward from that starting point.

6. Insert the three fully assembled test tubes in the test tube rack, and make sure to rotate the test tube rack so that the tubes face the light.

7. Record your starting time and initial observations in **Table 7.2**. Check your test tubes every 15 minutes to make note of the position of the phenol red solution in the volumeter relative to the starting point (in mm) and color of the phenol red solution. It will be helpful to hold a white sheet of paper behind the test tubes when assessing color of the phenol red solution.

8. After 60 minutes has elapsed you can make your final observations, turn off the lamp, and you can begin cleaning up. Return the phenol red solution to the beaker to be re-used and follow any other clean up instructions given by your instructor.

9. Copy the data down from another group near you that had the other color of light bulb onto **Table 7.3.** Answer the questions.

Light color:_____

Table 7.2: Data recorded from the photosynthesis and respiration experiment.

Time	Test Tube with *Elodea*		Test Tube with *Elodea* and Foil		Test Tube with no *Elodea*
	color	volumeter (mm)	color	volumeter (mm)	
Start					
15					
30					
45					
60					

Light color:_____

Table 7.3: Data recorded from the photosynthesis and respiration experiment.

Time	Test Tube with *Elodea*		Test Tube with *Elodea* and Foil		Test Tube with no *Elodea*
	color	volumeter (mm)	color	volumeter (mm)	
Start					
15					
30					
45					
60					

QUESTIONS:

1. How did the color of the tube containing the *Elodea* and not covered by foil change over the time of the experiment? How did this compare to the tube covered in foil? If anything happened explain why.

2. What happened to the water level in the volumeter in the tube containing the *Elodea* not covered by foil? How did this compare to the tube covered by foil?

3. What caused any movement of the water in the volumeter?

4. Were the predictions you made supported by your data? Explain your response.?

5. Why did we have a treatment group where the *Elodea* was covered by foil?

6. What was the purpose of the empty tube containing only the phenol red solution?

7. How did your results compare to the results with the other light bulb? Compare the results of the green vs. white light bulb treatments.

Chapter 7 Post Lab Questions

1. What is light and why are objects different colors?

2. How does light behave?

3. Where does photosynthesis take place in plants?

4. What is the relationship between photosynthesis and respiration?

5. What are pigments in general? What are accessory pigments?

6. What is an Rf value and why is it useful?

7. What was the source of carbon for the *Elodea* leaves?

8. If you placed an *Elodea* plant in a test tube containing reddish-pink phenol red solution and then covered it with aluminum foil, what color would you expect the solution to be later on? Why?

9. How do you explain why the water in the volumeter moved in response to the different treatments in the photosynthesis experiment?

Notes

Notes

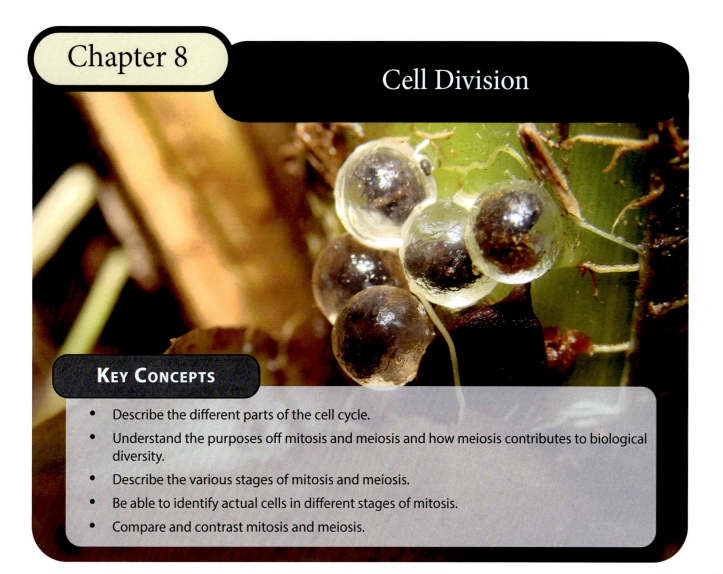

Chapter 8

Cell Division

KEY CONCEPTS

- Describe the different parts of the cell cycle.
- Understand the purposes off mitosis and meiosis and how meiosis contributes to biological diversity.
- Describe the various stages of mitosis and meiosis.
- Be able to identify actual cells in different stages of mitosis.
- Compare and contrast mitosis and meiosis.

Introduction

In the heart of a vibrant tropical forest, where the air is thick with humidity and the canopy above teems with life, a remarkable process unfolds in the damp leaf litter below. Among the myriad of life forms, the tiny eggs of a frog species known for direct development rest hidden. Unlike many of their amphibian relatives, these frogs skip the vulnerable tadpole stage. Instead, within each delicate egg, a complex and astonishing journey of cell division and differentiation is underway. Cells rapidly multiply and specialize, orchestrating the intricate dance of embryogenesis that transforms a single fertilized cell into a fully formed, miniature frog. This direct development exemplifies the fundamental principles of cell division that underlie all forms of life, from the simplest organisms to the most complex. As the tropical forest thrives with energy and growth, so too do these hidden embryos.

To understand the development of these frog embryos, it's essential to explore the processes of mitosis and meiosis. Mitosis is the process by which a single cell divides to produce two identical diploid cells. Diploid cells contain two sets of chromosomes, one from each parent. This process is how the embryo grows, with cells dividing and multiplying to form tissues and organs. Mitosis also plays a vital role in healing our tissues after an injury. Meiosis, on the other hand, is responsible for producing gametes which are haploid reproductive cells (i.e., sperm and eggs). Haploid cells have one full set of chromosomes, or half the number of chromosomes compared to diploid cells. When a sperm and an egg unite during fertilization, they form a diploid developing zygote (early embryo). In the case of our tropical frog, the zygote undergoes numerous rounds of mitosis, allowing the embryo to develop directly into a froglet without passing through the tadpole stage.

Both mitosis and meiosis are preceded by a crucial phase called interphase. During interphase, the cell prepares for division by growing and replicating its DNA. This phase is divided into three stages: G1, S, and G2 (**Figure 8.1**). In the G1 phase, the cell grows and carries out normal functions. The S phase is when the cell's DNA is replicated, ensuring that each new cell will have a complete set of chromosomes. Finally, in the G2 phase, the cell continues to grow and makes final preparations for division. Although most cells will divide in the course of an organism's life, there are some that will stop dividing and enter a phase called "G0". Interphase is essential because it sets the stage for the accurate and orderly distribution of chromosomes during cell division. Cell division, whether mitosis or meiosis, happens in what is called the "M phase" of the cell cycle. In our tropical frog embryo, interphase ensures that each new cell formed during mitosis has the genetic information needed to develop properly.

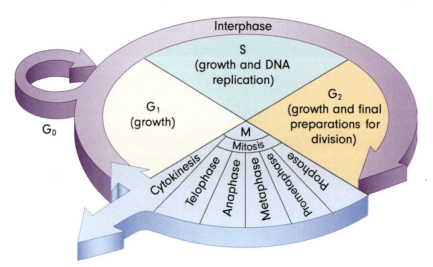

Figure 8.1: The cell cycle showing the different parts of interphase along with M phase also showing the stages of mitosis.

This interplay between meiosis and mitosis exemplifies one of the tenants of The Cell Theory, that all cells arise from preexisting cells. In this lab we will be exploring mitosis and meiosis with various exercises.

8.1 Mitosis copies cells

Mitosis is a fundamental process of cell division that ensures the growth, renewal, and repair of multicellular organisms. It involves a single cell dividing once to produce two genetically identical cells called **daughter cells**. Each daughter cell contains the same number of chromosomes as the original cell. This process allows organisms to grow, develop, and heal by creating new cells that are exact copies of the original. Mitosis maintains genetic continuity, ensuring that each new cell has the same genetic information as its predecessor. By producing identical diploid cells, mitosis plays a crucial role in tissue maintenance and regeneration, highlighting its importance in the life cycle of all multicellular organisms.

Mitosis is divided into six stages (**Figure 8.2**). However, it is important to mention that the process itself occurs fluidly and continuously. The stages below serve as a way to understand and describe the key events that happen during this intricate process. By breaking down mitosis into these stages, we can better grasp how cells ensure accurate distribution of their replicated genetic material.

- **Prophase**: Chromosomes condense and become visible under a microscope. Recall that the DNA was replicated during interphase. Replicated DNA double helices from the same chromosome are joined together at a location toward the center called the centromere. When joined together, each replicated chromosome is called a **sister chromatid**. The nuclear envelope begins to break down, and the mitotic spindle starts to form from the centrosomes, which move toward opposite poles of the cell. The **centrosome** is the organelle containing the two centrioles in animal cells from which the spindle fibers arise that separate the chromosomes during cell division. The spindle fibers are made of microtubules, which are part of the cytoskeleton. Plants do not have centrosomes with centrioles to organize the spindle fibers.

- **Prometaphase**: The nuclear envelope completely disintegrates, allowing spindle fibers to attach to special proteins around the centromeres called kinetochore proteins. The chromosomes continue to condense, and the mitotic spindle is fully formed. Toward the end of this phase, the spindle fibers begin to move the chromosomes toward the center of the cell.

- **Metaphase**: The replicated chromosomes line up at the cell's equatorial plane, known as the metaphase plate.

- **Anaphase**: The sister chromatids are separated and pulled toward opposite poles of the cell by the spindle fibers. This ensures that each new cell will receive an identical set of chromosomes.

- **Telophase**: The chromosomes arrive at the poles and begin to decondense. The nuclear envelope re-forms around each set of chromosomes, resulting in two distinct nuclei within the cell. Spindle fibers and the centrioles they come from begin to disappear.

- **Cytokinesis**: This final stage is actually separate from mitosis and begins during telophase. It involves the division of the cytoplasm, resulting in two completely separate daughter cells. In animal cells, a cleavage furrow forms to split the cell, while in plant cells, a cell plate forms to divide the cell (**Figure 8.3**).

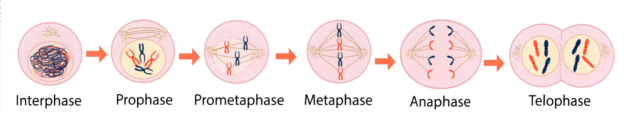

Interphase   Prophase   Prometaphase   Metaphase   Anaphase   Telophase

Figure 8.2: The stages of mitosis outlined in an animal cell.

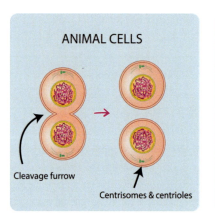
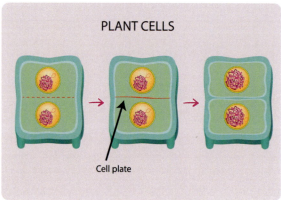

ANIMAL CELLS

Cleavage furrow

Centrisomes & centrioles

PLANT CELLS

Cell plate

Figure 8.3: The differences between plant and animal cells in cytokinesis.

EXERCISE: LEARN MITOSIS WITH POP BEADS

We are going to now explore the process of mitosis using pop beads to replicate the various stages. The pop beads will represent the chromosomes that will be moving around during the division process. We can use each bead to represent a chunk of DNA that makes up the condensed chromosome we can see beginning in prophase. Centromeres will be represented in this exercise by magnetic connection points.

SET-UP

We will begin this exercise by assembling our chromosomes. Groups of organisms have different numbers of chromosomes that vary in length. Each individual has two of each type of chromosome (which is the standard diploid condition). For example, in each of our cells we have 23 chromosome types. One set of 23 chromosomes comes from mom and another set of 23 chromosomes comes from dad. This means you have a total of 46 chromosomes (two of each of the 23 types of chromosomes). To simplify life and avoid an 8 hour lab, we will only follow two homologous pairs of chromosomes, which amounts to four total chromosomes, through the process of mitosis (and later meiosis). Because the chromosomes are replicated prior to mitosis, this means we will need to construct a total of 8 sister chromatids for our exercise.

To assemble the eight chromosomes/chromatids we will need for our exercise, you will need to get 8 magnetic centromeres. We will attach pop beads to each side of each of the magnetic centromeres to make our chromosomes (**Figure 8.4**). We will have two types of chromosomes- a short chromosome and a long chromosome.

Construct the short chromosome by attaching two pop beads at each end of the centromere. You will need four of these (16 beads). Make half of these one color to represent one parent and use a second color for the other half to represent the other parent from which the chromosomes came.

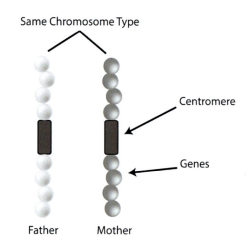

Same Chromosome Type

Centromere

Genes

Father   Mother

Figure 8.4: Example of how to assemble the pop bead chromosomes. This also depicts a homologous set of unreplicated chromosomes

Construct the long chromosome by attaching four beads to each side of the four magnetic centromeres for this (32 beads). Again, use one color bead for half of them and the second color for the other half. After you finish setting up these chromosomes you are ready to begin the rest of the exercise.

ACTIVITY INSTRUCTIONS

1. We will start the mitosis activity by pretending that we have our original 4 chromosomes (two of each type) and that our cell is in the early part of interphase (G1). The first thing we will do in preparation for mitosis will be to replicate the DNA in the S phase of interphase. To do this, simply add the identical sets of chromosomes to the four we are starting with and let the magnetic centromeres connect (**see example in Figure 8.5**).

2. Now we are ready to enter prophase after finishing interphase. This is where we can officially start to model the process of mitosis. Use the pop bead chromosomes to take your hypothetical cells through the entire process in a step-by-step manner. Make sure you go through all of the phases, as depicted in **Figure 8.2**. If you are working with a partner, make sure that both of you take turns doing this and explain each step as you go.

3. In the space below, **draw** what your pop bead cell looks like during metaphase.

Replicated Chromosomes

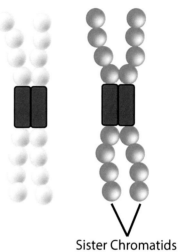

Sister Chromatids

Figure 8.5: Example of the replicated chromosomes held together at the centromere regions

8.2 Meiosis is for producing gametes

Meiosis is a specialized type of cell division that reduces the chromosome number by half, resulting in the production of haploid gametes, such as sperm and eggs. Taking the original cell from a diploid to a haploid condition is accomplished through two consecutive cell divisions—meiosis I and meiosis II. Unlike mitosis, which creates identical diploid cells, meiosis results in genetically unique cells that contribute to the maintenance of genetic diversity. This is achieved through swapping identical genetic regions between homologous chromosomes.

The genetic diversity resulting from sexual reproduction offers numerous benefits to populations and species. By combining genetic material from two parents, sexual reproduction creates offspring with unique genetic combinations. This diversity enhances the ability of populations to adapt to changing environments, resist diseases, and survive under various ecological conditions. Ultimately, the genetic diversity achieved through sexual reproduction strengthens the resilience and adaptability of species, ensuring their long-term survival and evolutionary potential.

Many of the processes that lead to genetic diversity in meiosis happen in the first part of meiosis—Meiosis I. During this stage, homologous chromosomes pair up in a process called **synapsis**. **Homologous chromosomes** are pairs of the same type of chromosomes, one from the mother and one from the father, containing the same genes but potentially different alleles. The homologous chromosomes physically join during synapsis to form structures known as **tetrads**, which consist of four chromatids. This close pairing allows for **crossing over**, where segments of genetic material are exchanged between homologous chromosomes (**Figure 8.6**). This recombination creates new combinations of alleles, significantly contributing to genetic variation. Additionally, during metaphase I, the way these chromosome pairs line up along the metaphase plate is random, leading to independent assortment. This means that the resulting gametes have a mix of maternal and paternal chromosomes, further enhancing genetic diversity. These key events in Meiosis I are fundamental to producing the variation that drives evolution and adaptation in populations.

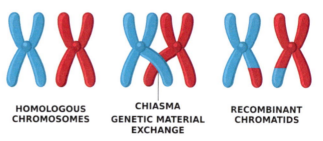

HOMOLOGOUS CHROMOSOMES

CHIASMA GENETIC MATERIAL EXCHANGE

RECOMBINANT CHROMATIDS

Figure 8.6: The process of homologous chromosomes pairing up and exchanging a given DNA segment during crossing over

STEPS OF MEIOSIS I (Figure 8.7)

- **Prophase I**: Chromosomes condense and become visible. Homologous chromosomes pair up in a process called synapsis. The paired homologous chromosomes form tetrads, each consisting of four chromatids. Crossing over occurs, where segments of genetic material are exchanged between the non-sister chromatids of homologous chromosomes. The nuclear envelope breaks down, and the mitotic spindle begins to form from the centrosomes, which move to opposite poles of the cell.

- **Metaphase I**: Tetrads align at the cell's equatorial plane, known as the metaphase plate. Spindle fibers attach to the centromeres of the homologous chromosomes. The alignment of tetrads is random, leading to independent assortment.

- **Anaphase I**: Homologous chromosomes are pulled apart to opposite poles of the cell by the spindle fibers. Sister chromatids remain attached at their centromeres.

- **Telophase I**: Chromosomes arrive at the poles of the cell. The nuclear envelope re-forms around each set of chromosomes. Cytokinesis begins, dividing the cytoplasm to form two **haploid** cells.

Meiosis I

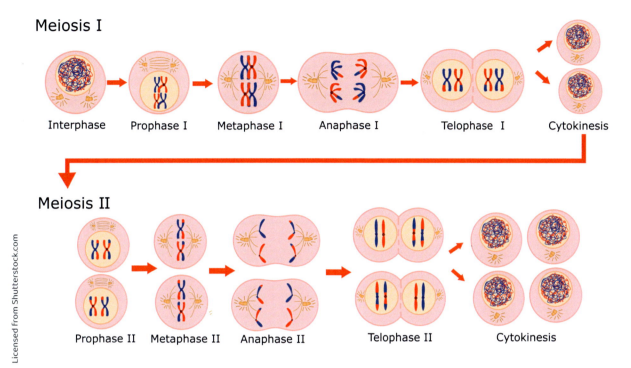

Meiosis II

Figure 8.7: The steps of both meiosis I and meiosis II that result in four haploid cells

By comparison to meiosis I, the process of meiosis II is straightforward. Meiosis II resembles a normal mitotic division. The two haploid cells produced in meiosis I undergo a second division without further DNA replication. This division separates the sister chromatids, resulting in four genetically unique haploid cells. Each of these cells contains a single set of chromosomes, ready to contribute to genetic diversity through fertilization. Meiosis II ensures that the chromosome number remains consistent across generations while still promoting genetic variation.

STEPS OF MEIOSIS II (Figure 8.7)

- **Prophase II**: Chromosomes condense and become visible again in sister chromatid form. The nuclear envelope breaks down again after it reformed briefly, and the spindle apparatus forms as centrosomes move to opposite poles of the cell.

- **Metaphase II**: Spindle fibers from opposite poles have attached to the centromeres of each chromosome to move them toward the center. Chromosomes align individually along the metaphase plate, similar to what happens in mitosis.

- **Anaphase II**: Sister chromatids are finally separated and pulled to opposite poles of the cell by the spindle fibers. Each chromatid is now considered an individual chromosome.

- **Telophase II**: Chromatids arrive at the poles, and the nuclear envelope re-forms around each set of chromosomes as they begin to unwind into chromatin. Cytokinesis begins, dividing the cytoplasm to form four genetically unique haploid cells.

After meiosis, the resulting haploid cells undergo further maturation through a process called **gametogenesis**. In males, this process is known as **spermatogenesis** and occurs in the testes, producing mature sperm cells. In females, it is called **oogenesis** and occurs in the ovaries, producing mature egg cells or ova. Gametogenesis ensures that each gamete is fully equipped for fertilization, carrying half the genetic material needed to form a new organism. This maturation process involves cellular differentiation and, in the case of oogenesis, the formation of one viable egg cell and polar bodies. Through gametogenesis, the haploid cells generated by meiosis become functional gametes capable of participating in sexual reproduction. When sperm and egg cells meet, they will fuse to form an initial single cell that is diploid called a **zygote**. After the zygote is formed, it will continue to divide in a systematic way through mitosis as the embryo develops.

EXERCISE: MODEL MEIOSIS WITH POP BEADS

Now that you are familiar with the pop beads from your mitosis exercise, we can branch out to the next level by using them to model the process of meiosis. Once again, the cells begin in interphase where they prepare for meiosis in several ways, including to replicate the DNA. The process of crossing over happens in the first part of meiosis. We can easily represent this using the pop beads by taking part of one sister chromatid and swapping it with the same part of the other chromatid- pop one section off and the new section on (**Figure 8.8**). It's important to always make sure the amount of genetic material exchanged is equal like it would be in real life.

Figure 8.8: Chromosomes made of pop beads crossing over while as a tetrad

Take your chromosomes through the process of meiosis. In the spaces below, draw out what your model cell with its chromosomes made of pop beads looks like at the different stages.

Prophase I

Metaphase I

Anaphase I

QUESTIONS ABOUT MEIOSIS I

1. At the end of meiosis I, how many chromosomes are in each of the two cells?

2. Are the chromosomes replicated or non-replicated at the end of meiosis I?

3. Are the resulting cells haploid or diploid?

Telophase I

Having finished with meiosis I and answered the questions, now proceed to take your bead chromosomes through the process of meiosis II. Do this for both cells that resulted from the first meiotic division. Remember, the purpose of meiosis II is to separate the sister chromatids, rather than the homologous pairs of chromosomes.

Prophase II

Metaphase II

Anaphase II

Telophase II

QUESTIONS ABOUT MEIOSIS II

1. From the work that you did, how many cells were produced at the end of meiosis II?

2. Were the cells produced at the end of meiosis II diploid or haploid?

3. Are there any of the four cells that are genetically identical to each other?

Comparing and contrasting the stages and overall processes of meiosis and mitosis is essential for a comprehensive understanding of each. This comparison highlights both their similarities and differences. The following questions are designed to guide you in examining these processes in detail.

1. Compare what happens in prophase of meiosis to prophase 1 in meiosis. What is similar and what is different?

2. Compare metaphase in mitosis to what happens in metaphase in meiosis I. How does it compare to metaphase in meiosis II?

3. Compare what happens during anaphase in mitosis to anaphase I in meiosis.

4. How does anaphase I in meiosis compare to anaphase II in meiosis?

5. Compare what happens in telophase during mitosis to what happens in meiosis.

8.3 Examining mitosis in real cells

Mitosis plays a crucial role in the expansion of a plant's root system. As the plant grows, cells in the root meristem, a region of actively dividing cells near the root tip, undergo mitosis. This process produces new cells that differentiate into various types of root tissues, such as the epidermis, cortex, and vascular tissues. These new cells allow the root to extend deeper into the soil, enhancing the plant's ability to absorb water and essential nutrients. The continuous production of new cells through mitosis also aids in the formation of root hairs, which increase the surface area for absorption. Additionally, mitosis is vital for the repair and replacement of damaged root cells, ensuring the root system remains healthy and functional.

The root tip is organized into distinct zones (**Figure 8.9**), each playing a specific role in root growth and development. At the very tip is the **root cap**, which is a protective layer of cells that shields the delicate mitotic tissue behind it as the root pushes through the soil. Just behind the root cap is the **zone of cell division**, where cells in the root meristem actively undergo mitosis to produce new cells that contribute to root growth. Following this is the **zone of elongation**, where newly formed cells increase in size to push the root further into the soil. Finally, there is the **zone of maturation**, where cells complete their differentiation into various specialized cell types, such as root hairs, which enhance the root's ability to absorb water and nutrients.

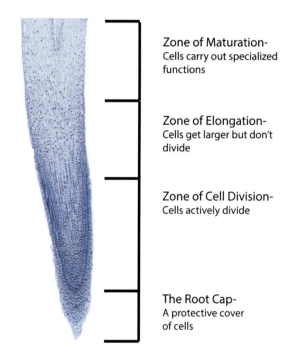

Zone of Maturation-
Cells carry out specialized functions

Zone of Elongation-
Cells get larger but don't divide

Zone of Cell Division-
Cells actively divide

The Root Cap-
A protective cover of cells

Figure 8.9: Zones of an onion root tip

EXERCISE: IDENTIFY MITOSiS STAGES IN ONION ROOT TIPS

In this exercise, you will examine pre-mounted slices of an onion root tip to identify the stages of mitosis in cells located in the zone of cell division. Observing the number of cells in each stage will help us understand how long cells spend in each stage of mitosis, in addition to learning how to identify the stage of mitosis real cells are in. Some stages pass quickly, while others take longer. Each root tip on the slide serves as a snapshot in time; the more cells you find in a particular stage, the longer the cells spend in that stage. Note that it may sometimes be challenging to determine the exact stage due to the continuous nature of cell division. Use **Figure 8.10** as a reference.

| Interphase | Prophase | Metaphase | Anaphase | Telophase |

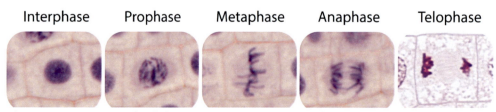

Figure 8.10: Stages of cell division in onion root tip cells

Retrieve a prepared slide of the onion root tip from the front of the room. Each slide contains two root tips. Use a compound light microscope to locate the zone of cell division in the root. Once you find the zone of cell division, systematically scan the area to count 100 cells and record how many are in each stage of mitosis. One partner should look through the microscope to identify the stages, while the other partner records the data as tick marks on a piece of paper. Record the count of cells in each stage in **Table 8.1**. After completing observations on the first root tip, move the slide to examine the second root tip. Switch roles with your partner to ensure both of you gain experience in identifying the stages of mitosis.

After you finish taking the data from both root tips on the slide, calculate the average number of cells in each stage for both root tips. Record these averages on the board where your instructor indicates. This will compile data from all groups in the class, creating a large sample size for more definitive conclusions about the duration of each stage of mitosis. You or your instructor will calculate average values from the class data and you can record those values in **Table 8.2**.

Table 8.1: The number of cells found in different stages of mitosis recorded from your work

Band color	Root 1	Root 2	Average
Interphase			
Prophase			
Metaphase			
Anaphase			
Telophase			

Space for tabulating the number of cells in each root tip:

Table 8.2: Class summary data for mitotic stages in onion root tips.

Band color	Class total for the number of cells	% of total
Interphase		
Prophase		
Metaphase		
Anaphase		
Telophase		
Total		

QUESTIONS

1. in which of the stage of mitosis did you find the most cells?

2. What can we say about a stage of mitosis if we find relatively few cells in that stage compared to the other stages?

3. Which stage of mitosis is the longest?

4. Which stage of mitosis is the shortest?

EXERCISE: FINDING STAGES OF MITOSIS IN FISH EMBRYOS

When sperm and egg meet, the egg is fertilized and the resulting single cell is called a **zygote** (**Figure 8.11**). This single diploid cell marks the beginning of a new individual and carries all the genetic information necessary for development. The zygote undergoes multiple rounds of cell division and differentiation called **cleavage** to begin developing into a multicellular organism. The result of this cell division is a ball of cells called a **morula** that quickly becomes a larger ball of cells called a **blastula**.

Obtain a microscope slide of a developing whitefish embryo. The microscope slide contains blastulas that are mounted and stained. Rapid cell division in early embryonic development means that we should be able to spot the various stages of mitosis in these blastulas. Find at least one cell in each stage of mitosis. Also observe some of the differences that can be seen when comparing these animal cells to the onion root tip cells you were just working with.

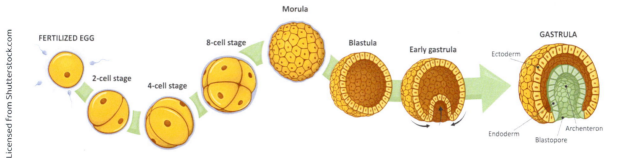

Figure 8.11: Stages early embryonic development following fertilization

1. Draw a picture in the square below of one of the blastula cells in anaphase of mitosis.

2. What are some of the differences between animal and plant cells in the cell division process, both apparent and less obvious?

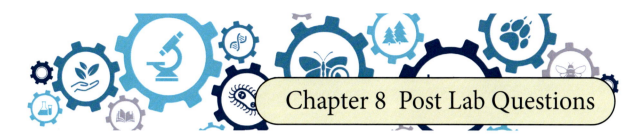

Chapter 8 Post Lab Questions

1. What is the general purpose of mitosis? Where do you see it in the human body?

2. What is the overall purpose of meiosis? What does it produce and where does it take place in the body?

3. What stage is the following cell in?

4. What is different between mitosis and meiosis I for the stage that the cell is in for question 3?

5. In your own words, describe what "homologous chromosomes" are and in what stage of meiosis do we find them?

6. How many divisions does a cell go through in mitosis vs. meiosis?

7. How many daughter cells are produced by mitosis and meiosis?

8. Below are short descriptions of something happening in mitosis or meiosis. Based on the short descriptions, name the stage (e.g., metaphase in mitosis or anaphase II in meiosis, etc). You can name stages in both mitosis or meiosis if it applies.

    a. Cytoplasm divides

    b. The tetrad structure forms

    c. The nuclear envelope dissolves/disappears.

    d. Sister chromatids separate toward opposite poles.

    e. Replicated chromosomes line up along the equator.

    f. Spindle fibers appear.

9. Describe at least 2 ways that meiosis contributes to genetic diversity.

Notes

Notes

Chapter 9

Heredity and Genetics

KEY CONCEPTS

- Be able to define and apply basic genetic terminology describing phenotypes and genotypes
- Understand how to construct a Punnett square for both single and two character crosses
- Analyze data from crosses to determine whether a trait is dominant or recessive.

Introduction

Before Gregor Mendel's groundbreaking work, the understanding of heredity was vague and often based on misconceptions. Many believed in the idea of blending inheritance, where offspring were thought to be a simple mix of parental traits, much like blending paints. Others subscribed to the concept of the inheritance of acquired characteristics, proposed by Jean-Baptiste Lamarck, which suggested that traits developed during an organism's lifetime could be passed on to its offspring. Gregor Mendel, an Austrian monk, revolutionized the field in the mid-19th century through his meticulous experiments with pea plants. By crossbreeding plants with distinct traits and analyzing the patterns of inheritance over several generations, Mendel discovered that traits are passed down through discrete units, now known as genes, laying the foundation for the field of genetics. His work demonstrated that some traits are inherited in specific ratios, rather than being blended, leading to the formulation of his laws of inheritance. Through his meticulous experiments, Mendel actually managed to figure out what happened during meiosis without ever looking at cells through a microscope and understood the ramifications of that process for how traits are passed along. His laws of segregation and independent assortment revealed predictable patterns of inheritance, transforming our understanding of biology and heredity.

Mendel's legacy from 1866 is not only of historical importance. He invented methods for tracking the inheritance of traits through generations and elucidating the genotypes of organisms that are still used by biologists today. For example, researchers in Finland have applied Mendel's methods to study the wood tiger moth (*Arctia plantaginis*- pictured above), which exhibits variable coloration in its forewings and hindwings. By tracking tens of thousands of individuals, they have learned how many chromosomes and genes are involved in determining their coloration and how different alleles of these genes interact to produce the diverse color variations seen in wild populations.

Any observable or measurable characteristic of an organism, whether visible or internal, is referred to as a **phenotype**. Phenotypes encompass conspicuous external features, such as coloration or limb structure, as well as internal aspects like organ function or metabolic processes. The **characteristic** is the aspect of the phenotype that varies and the **trait** is a particular form within that variation. For instance, in the wood tiger moth, wing color would be one of the characteristics and yellow wings would be a particular trait. These observable traits are ultimately determined by the organism's **genotype**, which is the specific gene or set of genes responsible for coding and regulating the trait. A particular gene can exist in several different forms or versions within a population, known as **alleles**. The interaction of alleles results in the expression of the phenotype, illustrating the crucial link between an organism's genetic makeup and its physical or functional characteristics.

The vast majority of living organisms are **diploid**, meaning they have two copies of every type of chromosome, and therefore every gene—with one set of chromosomes and genes inherited from each parent. This dual inheritance means that phenotypes are products of the interactions between the alleles for the same genes that exist across the two sets of chromosomes. In some cases, the two alleles are identical. Individuals with identical alleles are termed **homozygous**. Conversely, when each chromosome carries a different version of the gene, the individual is said to be **heterozygous**.

In genetics, dominance plays a crucial role in determining phenotypes. A **dominant allele** can mask the presence of a recessive allele, meaning the dominant allele solely determines the phenotype, even if only one dominant allele is present. **Recessive alleles**, on the other hand, require two copies—one from each parent—to be expressed in the phenotype that they code for.

We can combine terms for dominance and whether alleles are alike or different to describe genotypes using three key terms: homozygous dominant, homozygous recessive, and heterozygous (**Figure 9.1**). A **homozygous dominant** individual has two dominant alleles, resulting in the expression of the dominant trait. A **homozygous recessive** individual has two recessive alleles, leading to the expression of the recessive trait. A **heterozygous** individual possesses one dominant and one recessive allele, with the dominant allele typically determining the phenotype.

The typical gene is thousands of bases long and alleles of a gene can differ by one base. So, it does not make sense to write out the whole DNA sequence to effectively track genotypes made up of two alleles. Fortunately, there is a shorthand way of writing the genotype of an individual for a particular characteristic. The most basic way of doing this is to let a single letter represent a gene, for instance the letter "P" for the flower color of a pea plant. We can then use the uppercase form of the letter "P" to represent the dominant allele and the lowercase form to stand for the recessive allele. In pea plants, the allele for purple flowers is dominant (P) and the allele for white flowers is recessive (p). Now we have a way to effectively record the genotype of an individual pea plant for the flower color characteristic, where it is easy for us to interpret the phenotype at the same time. A plant with the genotype PP (homozygous dominant) or Pp (heterozygous) would

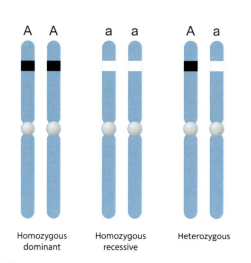

Homozygous dominant   Homozygous recessive   Heterozygous

Figure 9.1: Homologous chromosomes showing different genotypes.

have purple flowers. Only plants with the genotype pp (homozygous recessive) would exhibit white flowers. Any letter can be used to represent the genotype. In this case, we used "P" because the dominant allele gives us a purple flower. This notation system simplifies the representation and understanding of genetic inheritance patterns. A slightly different version of this basic notation can be used when we have more complex situations like multiple alleles in a population, rather than just two.

Genetics is often studied using **model organisms**—species extensively researched due to their unique advantages, such as short generation times, ease of maintenance, and well-mapped genomes. These organisms, including fruit flies (*Drosophila melanogaster*), mice (*Mus musculus*), and the nematode worm (*Caenorhabditis elegans*), serve as valuable tools for understanding fundamental biological processes. In the age of genomics, more organisms are becoming model organisms, allowing biologists to determine more detailed mechanisms for how traits are inherited. Insights gained from model organisms often translate to broader applications, helping scientists uncover genetic mechanisms conserved across species, including humans. In this lab, you will be examining patterns of inheritance in corn seed characteristics to identify dominant and recessive traits.

9.1 Single character crosses

Now that you have a basic understanding of the history of studying heredity and the terminology to describe genotypes and phenotypes, let's look closely at how we study the heredity patterns of single characteristics. Examining one characteristic and how it was passed from one generation to the next ultimately helped Gregor Mendel deduce the Law of Segregation. Let's look at an example from his work on pea plants.

To understand how a single characteristic was passed down, Mendel first created true breeding lines of parents that would only produce one of two traits. We will look at pea flower color, which is either white or purple. Once a group of plants was selectively bred to produce only white flowers and another group to produce only purple flowers, Mendel proceeded to the next step (**Figure 9.2**). He crossed the true breeding purple flower plants with the true breeding white flower plants. To help us understand the phenotypes and genotypes involved, we write the genotype of each parent being crossed. In this case, we cross a homozygous dominant (PP) parent with a homozygous recessive (pp) parent—so PP x pp. Today, we have a tool to help us visualize the expected results of a mating called a Punnet square, which was developed in 1905 by Reginald Punnett (after Mendel's death).

A **Punnett square** is a depiction of a cross between two parents, showing the likely genotypes and phenotypes of the resulting offspring. To construct a Punnett square, we take the genotype of one parent and place it along the top, and the other parent's genotype along the side. We split the genotype of each parent, separating the alleles to represent what would be passed down through a gamete. A set of four squares is formed when we bifurcate the genotypes with lines. We then carry the alleles over and down into the four squares. This shows us all possible genotypes that will result from the cross. In our case, PP x pp will yield only one genotype in all four squares: Pp (a heterozygote), which is purple. Mendel was puzzled by the disappearance of the white flower trait in the F1 offspring, but the Punnett square helps us see that the allele coding for the white flower (p) was masked by the dominant purple flower allele (P).

Mendel set up another cross to investigate further, this time using only the F1 generation, which consisted of all purple-flowered plants. The starting generation in any study of heredity is the **P generation** (for parent) and each successive generation is an F generation. The first generation from the parents is the **F1 generation**. When we set up the second cross (**Figure 9.2**), we see that the genotypes of both parents will be the same—Pp x Pp. This type of cross is called a "**monohybrid cross**" because both parents are heterozygous. To fill out a Punnett square for this monohybrid cross involving Mendel's pea plant flower color, we start by drawing a 2x2 grid. We label the top with one

parent's alleles (P and p) and the left side with the other parent's alleles (P and p). Each box represents a potential genotype of the offspring. The top left box receives a P from the top and a P from the side, resulting in PP. The top right box combines P from the top and p from the side, resulting in Pp. The bottom left box combines p from the top and P from the side, also resulting in Pp. Finally, the bottom right box combines p from both the top and the side, resulting in pp. The completed Punnett square shows one PP, two Pp, and one pp, corresponding to an expected phenotypic ratio of three purple-flowered plants to one white-flowered plant in the F2 generation, demonstrating Mendel's 3:1 ratio for a monohybrid cross.

The importance of the monohybrid cross between heterozygous pea plants (Pp x Pp) to Mendel's elucidation of his Law of Segregation lies in the predictable 3:1 phenotypic ratio of the offspring. By observing this ratio, Mendel deduced that each parent carries two alleles for a given trait, which separate or "segregate" during the formation of gametes (sperm and egg cells). This segregation ensures that each gamete receives only one allele from each parent. Upon fertilization, the offspring inherit one allele from each parent, restoring the diploid state. Mendel's careful counting and statistical analysis of the offspring from these crosses allowed him to formulate the Law of Segregation, which states that allele pairs separate during gamete formation (meiosis) and randomly unite at fertilization.

If we were doing this experimental cross with an actual group of pea plants, we would get more than just four seeds/offspring from that cross. Pea plants typically produce between 20 and 100 offspring or more per cross. Mendel and researchers today, like those working with the wood tiger moth, cross dozens to hundreds of individuals to obtain a large number of offspring and more accurately capture patterns of heredity. The Punnett square allows us to predict the proportion of those offspring that would exhibit particular genotypes and phenotypes.

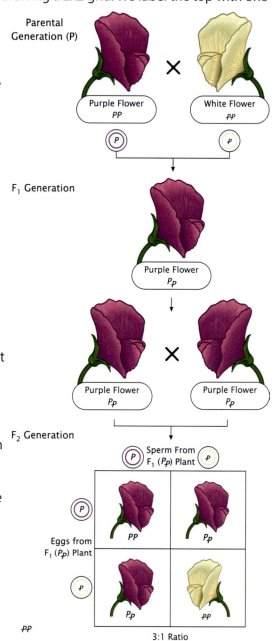

Figure 9.2: The crossing of true breeding purple and white flower parents to produce the F1 generation, followed by a monohybrid cross to produce the F2 generation.

EXERCISE: UNDERSTANDING MENDELIAN GENETICS

Answer the following questions:

1. One of the traits that Mendel looked at in the pea plants was the seed color. Some peas produced yellow seeds (Y) and other peas produced green seeds (y). If we were experimenting with pea color and we crossed a parent that was heterozygous with another that was homozygous recessive what would the resulting genotypes and phenotypes be?

2. If we crossed two heterozygous individuals with respect to pea seed color, what would the resulting genotypes and phenotypes be? Let's say there are 173 offspring resulting from that cross (each pea is a new individual).

    a. How many of the offspring would be green?

    b. How many offspring would be yellow?

    c. What percent of the offspring will be homozygous dominant?

    d. How many offspring are expected to be heterozygous?

9.2 Single characters: Assessing allele dominance in corn

Corn (*Zea mays*) is one of the most iconic vegetables, not only for its central role in diets and its association with the fall season, but also for its vast agricultural importance as a versatile and widely cultivated crop used for food, animal feed, biofuel, and numerous industrial products. Corn traces its origins to southern Mexico where it was domesticated by indigenous peoples around 9,000 years ago. Today, it stands as one of the world's most vital staple crops, extensively cultivated across the Americas, Africa, and Asia. Corn features a unique reproductive system with separate male and female flowers on the same plant. The male flowers form the tassel at the plant's apex, while the female flowers develop into ears along the stalk. Remarkably, each kernel on an ear of corn results from an individual fertilization event, making each kernel a genetically unique entity.

Kernel color and texture in corn are determined by specific genetic traits. The **aleurone** layer, a single cell layer just beneath the outer **pericarp** (seed coat), plays a key role in kernel color (**Figure 9.3**). In purple corn, the aleurone layer is rich in anthocyanin pigments, giving the kernels their distinctive purple hue. Conversely, yellow corn kernels lack anthocyanin pigments in the aleurone layer, allowing the yellow carotenoid pigments in the endosperm to dominate. Kernel texture is influenced by genes controlling starch production and deposition. Round kernels result from smooth starch deposits, whereas wrinkled kernels arise from a mutation that disrupts normal starch production, leading to a shrunken and wrinkled appearance. These traits are inherited according to Mendelian principles, making corn an excellent model for studying genetic inheritance patterns, much like Mendel's classic experiments with pea plants.

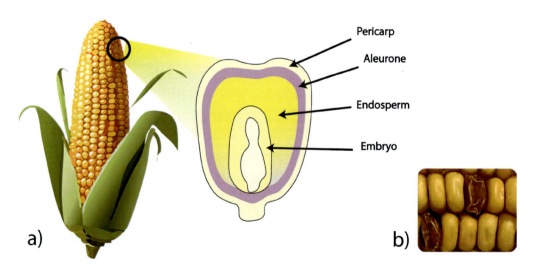

Figure 9.3: (a) The anatomy of a corn seed showing the layers that contribute to the color of the seed. (b) The difference between round and wrinkled seeds

EXERCISE: LOOKING AT SEED COLOR

This is the first of two exercises where we will be looking to determine which phenotype of the corn seed is dominant. First, we will focus on the seed's color that can be either purple or yellow. The gene that determines color is on chromosome 2. Work through the questions below to make your determination. For consistency, use the letter "P" in the genotype descriptions for seed color.

For color, there are two possibilities. The dominant color allele could code for yellow, or it could code for purple. The first thing we should do is to establish what we expect to find in either case. This is an important thing to do in any research situation- think about what the results look like for supporting and not supporting your hypothesis.

A. Possibility 1- The yellow allele is dominant.

1. What are the possible genotypes of the yellow kernels in this scenariio?

2. What are the possible genotypes of the purple kernels if the yellow allele is dominant?

3. Construct a Punnett square for a heterozygote crossed with a homozygous recessive individual and also construct one for two heterozygous individuals being crossed. What are the proportions of offspring phenotypes that we might expect from those scenarios?

B. Possibility 2- The purple allele is dominant.

1. What are the possible genotypes of the purple kernels in this scenariio?

2. What are the possible genotypes of the yellow kernels if the purple allele is dominant?

3. Construct a Punnett square for a heterozygote crossed with a homozygous recessive individual and also construct one for two heterozygous individuals being crossed. What are the proportions of offspring phenotypes that we might expect from those scenarios?

C. Take data from an ear of corn to see which of the two possibilities is correct.

Count at least 100 kernals of the appropriate ear of corn from the front of the room for testing what color is dominant. You can have one partner make tick marks for each kernel type and then add them up while the other partner calls out the kernel color.

Purple: _____

Yellow: _____

Total: _____

Space for counting:

QUESTIONS:

1. Figure out the ratio of purple to yellow kernels of corn.

2. What color do you think is dominant? Explain your reasoning.

3. What were the genotypes of the parents that produced this ear of corn?

EXERCISE: LOOKING AT SEED SHAPE

We are now ready to move to the next characteristic that varies across kernels of corn- seed shape. The gene for the endosperm composition that determines shape is on chromosome 4. Here, we will repeat most of the procedure we just carried out to look at seed color. There are two possibilities for seed shape. The seed could be round or wrinkled, depending on the starch inside the seed. Use the letter "R" to code the genotypes in this example to keep things the same.

A. Possibility 1- The round allele is dominant.

1. What are the possible genotypes of the round kernels in this scenario?

2. What are the possible genotypes of the wrinkled kernels if the round allele is dominant?

3. Construct a Punnett square for a heterozygote crossed with a homozygous recessive individual and also construct one for two heterozygous individuals being crossed. What are the proportions of offspring phenotypes that we might expect from those scenarios?

B. Possibility 2- The wrinkled allele is dominant.

1. What are the possible genotypes of the wrinkled kernels in this scenariio?

2. What are the possible genotypes of the round kernels if the wrinkled allele is dominant?

3. Construct a Punnett square for a heterozygote crossed with a homozygous recessive individual and also construct one for two heterozygous individuals being crossed. What are the proportions of offspring phenotypes that we might expect from those scenarios?

C. Take data from an ear of corn to see which of the two possibilities is correct.

Count at least 100 kernals of the appropriate ear of corn from the front of the room for testing what shape is dominant. You can have one partner make tick marks for each kernel type and then add them up while the other partner calls out the kernel color.

Wrinkled: _____

Round: _____

Total: _____

Space for counting:

QUESTIONS:

1. Figure out the ratio of wrinkled to round kernels of corn.

2. What shape do you think is dominant? Explain your reasoning.

3. What were the genotypes of the parents that produced this ear of corn?

9.3 Two character crosses

Gregor Mendel's pioneering work in genetics extended beyond the study of single characters. He also explored the inheritance patterns of two different characters simultaneously. For instance, he crossed pea plants that differed in two characters, such as seed color and seed shape. He would then observe the resulting phenotypic ratios in the offspring. By following two characters at the same time, Mendel discovered the Law of Independent Assortment, which states that alleles for different traits are distributed to gametes independently of one another. If genes are linked, rather than assorting independently, the patterns of inheritance for the individual characteristics will differ. Genes typically assort independently when they are on different chromosomes or far apart on the same chromosome. However, genes that are close together on the same chromosome tend to be linked and follow each other because they are not separated by crossing over during meiosis.

One of the techniques Mendel used, as he was following two characteristics, was the dihybrid cross. In a **dihybrid cross**, two parents that are both heterozygous for both characteristics of interest are crossed. He discovered that this cross resulted in a predictable phenotypic ratio, much like the 3:1 ratio observed in the monohybrid cross. This finding was crucial in advancing our understanding of genetic inheritance and the behavior of chromosomes during meiosis.

In this part of the lab, we will look at the seed color and shape of our corn seed. We will test whether particular ears of corn that vary in both of these characteristics are the product of a dihybrid cross.

EXERCISE: TESTING FOR DIHYBRID CORN

In the previous exercises, you determined which alleles for color and seed shape were dominant. Now, you will examine an ear of corn that was bred to show variability in both of these characteristics. The ear of corn you are studying is the product of a dihybrid cross. The parent's genotypes are PpRr x PpRr. You will compare the phenotypic ratios of the seeds in this ear of corn to the expected ratios from Mendel's work with peas for a dihybrid cross.

WHAT ARE OUR PREDICTIONS?

Step 1: The first thing we need to do is establish our expectations by determining the phenotypic ratios we should anticipate. To do this, we will set up a two-character cross and create a Punnett square. The first step in assembling the Punnett square is to determine the alleles present in the gametes– eggs and sperm. When examining single characters, this process is straightforward. Each gamete receives only one allele from the parent's genotype, according to the Law of Segregation. Here, the same principle applies, but with two characters, each gamete must include alleles for both traits. This means our Punnett square will need to represent two alleles in each gamete—one for each character.

List the different allele combinations found in the gametes of each parent (PpRr) below:

Step 2: Now it is time to set up the Punnett square. Write one parent's gametes across the top of the square you will make and write the other parent's gametes down the side. This is an identical process to how we set up a Punnett square for single character crosses, but with more allele combinations in our gametes, there are more than four squares.

Draw out the Punnett square in the space below and fill it in.

Step 3: We now have to determine what the phenotypes are to be able to figure out what the phenotypic ratios will be. To do this, look at each allele pair to determine the trait for seed color and seed shape for each genotype. Write those combinations below. There should be four phenotypic combinations.

Step 4: Time to figure out the phenotypic ratios. To do this, take the different phenotypic combinations and write the different genotypes that would produce those below the description. We need to then count how many of each of those genotypes appear in the Punnett square table that you created above.

GATHER THE DATA

Now that we have our expectations in place for what a dihybrid cross should produce, let's see if the real ears of corn match those expectations. To do this, you will count 200 kernels of corn from the ear of corn you obtained from the front of the room for this exercise and compare how many there are of each phenotype to the predicted ratio you figured out on the previous page.

Step 1: If you count 200 kernels of corn, how many do you expect there to be of each phenotype (use the ratios to calculate as a % of 200)? Record this information in the space below and in the "expected #" column in **Table 9.1** along with the ratio itself:

Phenotype 1:   Phenotype 2:

Phenotype 3:   Phenotype 4:

Step 2: Count the number of kernels that fall into each of the four phenotypic categories you determined in step 3 on the previous page. Use this space to make tick marks as you count. Record the data in the "count data" column in **Table 9.1.**

Phenotype 1:_____

Phenotype 2:_____

Phenotype 3:_____

Phenotype 4:_____

Step 3: Fill in **Table 9.1** below with your expectations and the data you collected. We will compare the numbers of kernels you expected to be of each particular phenotype to the number of kernels you actually found. If they are similar, then the ear of corn matches our expectations and the predictions that Mendel established.

Table 9.1: Comparison between the predicted number of kernels of each phenotype and the actual number of kernals we counted that matched each of the four phenotypes.

Phenotypes	Expected ratio	Expected #	Count data	Difference (#)

COMPARE THE RESULTS:

1. How do the numbers of kernels you predicted there to be for each phenotype compare to the number of kernels you found? Was there much of a difference?

2. Remembering our discussion in the background about what contributes to there being independent assortment, can you explain why these two characters behaved as such? You might need to refer back to the information about each characteristic in the single character corn work (Section 9.2).

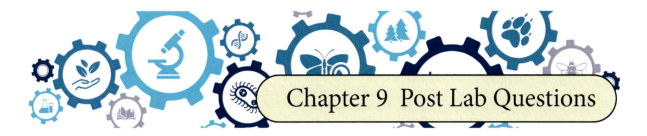

Chapter 9 Post Lab Questions

1. What is the genotype of an individual in general? How does that relate to the phenotype?

2. What is the difference between a dominant allele and a recessive allele?

3. What is the underlying cause of an individual kernel of corn being yellow or purple?

4. If a plant species can have red flowers or white flowers and white flowers are recessive, then what percent of offspring would be white in a cross between a white flowered plant and a heterozygous red flowered plant? Construct a Punnett square to answer the question in the space below.

5. Corn is not the only farm related entity that has interesting characteristics determined in a simple Mendelian inheritance fashion. Some species of chickens have feathers on their feet that also follow the patterns Mendel discovered. In these cases, feathered feet are dominant (F) and feet without feathers are a recessive trait (f). If I have a male with feathers on his feet (Ff) and cross him with a female hen with feathers on her feet (FF), then how many of the 10 offpsring will be expected to have feathers on their feet? Construct a Punnett square to answer this question below.

6. If the genes for corn seed color and corn seed shape were on the same chromosome, then would we expect the pattern of ineritance and the phenotypic ratios to change? Why or why not?

7. In chickens, the black feather color (B) is dominant over white feather color (b), and the single comb shape (S) is dominant over the pea comb shape (s). A chicken that is heterozygous for both traits (BbSs) is crossed with another chicken that has black feathers with a pea comb shape (Bbss).

Construct a Punnett square for this dihybrid cross.

Based on the Punnett square, determine the expected phenotypic ratio of the offspring.

Out of 12 offspring, how many would you expect to have black feathers and a single comb?

Notes

Notes

Chapter 10

Molecular Genetics

KEY CONCEPTS

- Know what the field of molecular genetics encompass and how it can be broadly applied to many different fields of biology
- Understand what a restriction enzyme is and what they can be used for.
- Be able to describe how to set up DNA gel electrophoresis and how it works.
- Understand how fragment size can be used to predict the banding pattern on a gel
- Interpret electrophoresis gel banding patterns.

Introduction

For centuries, humans have been captivated by the beauty and diversity of birds, meticulously observing their behaviors, songs, and migrations. Among these, the house sparrow (*Passer domesticus*) has thrived alongside human civilization, originating in the Middle East and spreading across Europe, Asia, and eventually worldwide through human activity. In only around 170 years, house sparrows managed to colonize the entire globe (just not Antarctica… yet). This close association with humans has provided a unique window into how urbanization affects wildlife. Molecular genetics has revolutionized our understanding of these resilient birds. Studies reveal significant genetic divergence between urban and rural populations, with urban sparrows exhibiting unique adaptations in metabolism, immunity, and behavior. For instance, urban sparrows show genetic changes that help them exploit diverse diets, cope with pollutants, and handle the stresses of city life. Research has found that urban sparrows have higher concentrations of lead and stress hormones, affecting their health and reproductive success. Such findings highlight how molecular techniques can uncover the impacts of human activities on wildlife.

Beyond sparrows, molecular genetics has expanded our understanding of avian biodiversity and evolution. Techniques like DNA barcoding and whole-genome sequencing have uncovered hidden/ cryptic species and redefined relationships among major groups, leading to the realization that there may be double the number of bird species we currently recognize. Genomic studies of Darwin's finches have revealed the genetic basis of beak shape variation, a key adaptation to different ecological niches. Similarly, research on European robins has provided insights into the genetic adaptations underlying their migratory patterns.

These advancements illustrate the power of molecular genetics to explore ecological and behavioral aspects of birds, transforming our approach to studying biodiversity. Beyond exploring various facets of biology, molecular genetics is also applied extensively in medicine, revolutionizing the field by enabling personalized treatments, advancing genetic therapies, and improving our understanding of disease mechanisms. Applied molecular genetics is crucial in agriculture, industry, and environmental management, forming part of the broader field of biotechnology, which seeks to apply science to solve practical problems. **Biotechnology** utilizes living organisms, cells, and biological systems to develop innovative products and technologies, such as genetically modified crops, biopharmaceuticals, biofuels, and diagnostic tools.

Many techniques and analyses from molecular genetics are used in both pure investigative research and applied research. In today's lab you will focus on learning how restriction enzymes and gel electrophoresis work to analyze similarities and difference among DNA samples. Although similar techniques are widely showcased on various TV crime dramas to determine who committed a crime, this same technology is used in a wide variety of applications – including conservation biology. Starting in the 1990's, the power of molecular biology for use in detecting illegal whale meat and other endangered wildlife-based products began to be recognized. We will be using restriction enzymes and gel electrophoresis today to analyze some possible wildlife contraband.

10.1 DNA extraction and amplification

Whale hunting whales is mostly illegal throughout the world, with notable exceptions in countries like Japan, Norway, and Iceland. Japan's whale hunting evolved from small-scale coastal whaling to commercial whaling in the 20th century. Despite a 1986 moratorium by the International Whaling Commission (IWC), which was established under the International Convention for the Regulation of Whaling (ICRW) treaty, Japan continued whaling for "scientific research," targeting minke, Bryde's, and sei whales. In 2019, Japan left the IWC and resumed commercial whaling within its exclusive economic zone, primarily harvesting these species. Protected species like blue, humpback, and right whales remain off-limits. Efforts to boost domestic whale meat consumption persist amid international controversy over sustainability and ethics. Occasionally, meat is smuggled from where it is legal in Japan to areas where it is illegal. Countries where this meat ends up include South Korea, China, and the United States on rare occasion. Once a whale is butchered it is almost impossible to differentiate it from other types of meat. Illegal whale is often sold as a type of large game fish, such as swordfish or another rarer species.

In today's lab we will pretend that we are running an independent genetic analysis lab. We recently received a sample of possible illegal whale meat obtained from a restaurant in Los Angeles, where illegal whale meat has been sold by restaurants in the past. Our task is to assess whether the meat sample is from a whale or fish, as it was marked where the sample was picked up. The first step in identifying what the source of the sample is to extract the DNA.

To extract DNA for genetic analyses, we start by breaking down the cells in our sample by using a lysis buffer containing detergents to dissolve the cell membranes and release the DNA. Next, we add a protease enzyme to remove proteins and other contaminants. We then spin the sample at high speed in a centrifuge to separate the DNA from the cellular debris that are forced to the bottom of the tube and form a pellet. We can then extract the DNA in the liquid, wash it to further remove any impurities and then store it in alcohol for the next step, which is amplification.

We will need to increase the amount of DNA to carry out some analyses. To amplify the DNA, we will use a process called polymerase chain reaction (PCR). This technique allows us to make millions of copies of a specific DNA segment. In our case, we are amplifying a 4,000 bp section of the mitochondrial DNA that contains the COI gene that is commonly used in population and species identification work. We first mix the extracted DNA with a solution containing primers that match and bind to our target sequence, DNA polymerase enzyme, and nucleotides that is called a master mix (**Figure 10.1**). The PCR process involves repeated cycles of heating to separate the DNA strands, cooling so the primers can bind to their complementary sequences, and then heating again to allow the DNA polymerase to synthesize new DNA strands (**Figure 10.2**). The amount of DNA doubles after each cycle. These cycles are repeated multiple times to exponentially amplify the target DNA, making it easier to analyze and identify whether the sample is from a whale or a fish.

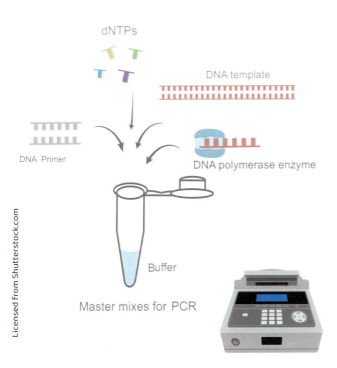

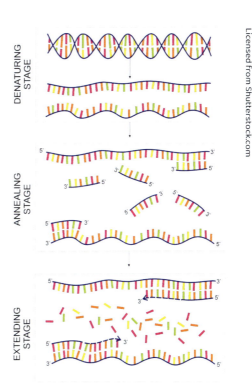

Figure 10.1: The mixture of primer, nucleotides, template DNA, and DNA polymerase that goes in with sample. The contents are then placed into the PCR machine (lower right corner) to amplify the DNA.

Figure 10.2: The sequence of steps in the process of PCR. Heating the mixture to separate the DNA, cooling to let the primers bind to the sequence where copying should begin, and heating to facilitate the DNA polymerases copying the DNA template.

10.2 Cutting DNA with restriction enzymes

The next step in the process of identifying whether our mystery meat is from a whale or fish is to use restriction enzymes to cut the DNA into fragments for analysis with gel electrophoresis. **Restriction enzymes**, also known as restriction endonucleases, are proteins that cut DNA at specific sequences called recognition sites. These enzymes can create two types of cuts: sticky ends or blunt ends (**Figure 10.3**). Sticky ends are created when the enzyme cuts the DNA in a staggered manner, leaving overhanging single-stranded sequences that can easily pair with complementary sequences. Blunt ends are produced when the enzyme cuts straight across both DNA strands, leaving no overhangs.

Choosing between sticky and blunt ends depends on the context of the experiment and has broad applications in biotechnology. Sticky ends, with their overhanging sequences, are ideal in some cases because they facilitate being able to take genetic information from one source and paste it into a new area of another chromosome. The sticky ends facilitate precise binding of the new DNA into a particular place. Blunt ends can be useful when an exact match is not needed to paste the new DNA into place. Blunt ends can also offer benefits in fragment analysis, depending on the specific context.

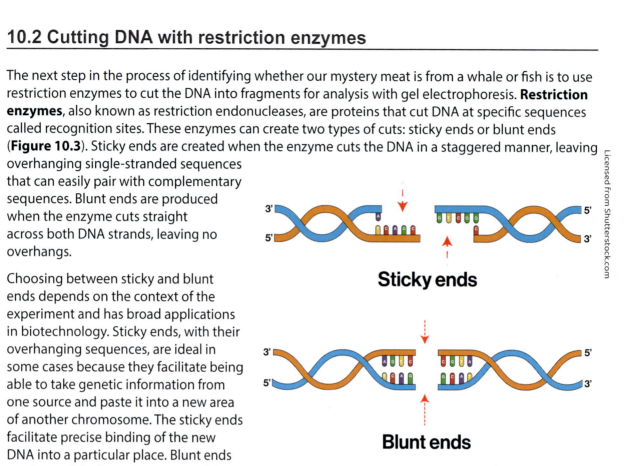

Figure 10.3: The difference between sticky ends and blunt ends. The partial single strands of the sticky ends can bind with other single straded DNA more easily.

The restriction enzyme EcoRI is commonly used in molecular biology for its ability to cut DNA at a specific palindromic sequence. The recognition site for EcoRI is 5'-GAATTC-3', and it cuts between the G and A nucleotides (**Figure 10.4**). This cut results in sticky ends with four-base overhangs: 5'-AATT-3'. When DNA is "digested" with EcoRI, it produces distinct fragments that can be separated and analyzed using techniques like gel electrophoresis, helping researchers study genetic variations and construct recombinant DNA molecules. If there are nucleotide differences between DNA molecules that affect the sequence of a restriction site, then restriction enzymes like EcoRI will produce different sized fragments in the different DNA molecules. In the exercise below, you will discover your inner enzyme and cut the two sample strands of DNA as the EcoRI enzyme would to see what differences there are in the number and size of the resulting fragments.

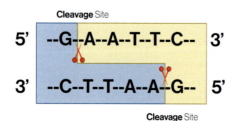

Figure 10.4: The cutting sites for the EcoRI enzyme when it encounters the palindromic sequence GAATTC.

EXERCISE: CUT DNA FRAGMENTS WITH THE ECORI ENZYME

In this exercise, we have three samples of DNA. One sample comes from a crime scene and the other two samples come from two different suspects. The EcoRI enzyme is added to the two sample to digest them, which means it will cut them at the target sequences as described. Because the sequence GAATTC is a palindrome, cuts are made on both strands of DNA. For simplicity, only one strand of the DNA sequence is shown below.

Draw lines where the EcoRI enzyme would cut the strands of DNA. List the number of fragments produced and the size of those fragments (i.e., count the bases).

SUSPECT A:

5' CAGTCAATTCCGGATTTTTCTAGAATTCAACGCCATATAATGCGCACGATTGGAATTAGGCC 3'

SUSPECT B:

5' CAGTGAATTCCGGATTTTTCTAGAATTCAACGCCATATAATGCGCACGATTGGAATTAGGCC 3'

CRIME SCENE

5' CAGTGAATTCCGGATTTTTCTAGAATTCAACGCCATATAATGCGCACGATTGGTATTAGGCC 3'

QUESTIONS:

1. Compare the three DNA sequences. In what two ways do they differ in general after being cut by the EcoRI restriction enzyme?

2. Can we use the EcoRI enzyme to tell the difference between the suspects?

3. Can we determine whether either of the suspects was at the crime scene?

10.3 Visualizing DNA fragments with gel electrophoresis

In the exercise from the previous section, we were able to see the DNA sequence and know exactly where the enzyme would cut, allowing us to easily count the number of bases (referred to as **base pairs** or bps) that made up the length of each fragment. However, in reality, sequencing the DNA samples is required to see the exact sequence of bases in a strand of DNA. Sequencing provides detailed information but is a costly and time-consuming process. A quicker and more affordable approach is broadly termed "fragment analysis," which allows us to determine the size of DNA fragments and visually distinguish between smaller and larger fragments without counting bases. We will use a type of fragment analysis called **Restriction Fragment Length Polymorphism** (**RFLP**) to determine whether our suspected whale meat is actually fish, as labeled, or whale. We already exposed our samples to restriction enzymes to produce fragments and are now ready to use gel electrophoresis to visualize the resulting fragments for easy analysis in this section.

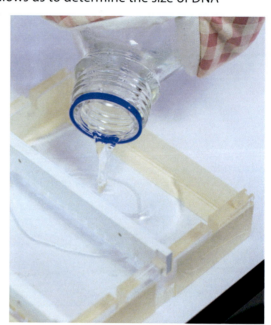

Figure 10.5: Pouring liquid agarose solution into a casting tray with two rows of gel combs to create 24 sample wells.

Gel electrophoresis allows us to separate the DNA fragments we produced with restriction enzymes by their size in a gel. The analysis begins with making the gel by dissolving **agarose** powder (a type of purified polysaccharide made from seaweed) into a heated buffer. Once dissolved and clear, the solution is poured into a **casting tray** (**Figure 10.5**). A gel comb is used to create **wells** in the gel, where we will later place our DNA samples. After the gel solidifies in the casting tray, it is moved into the **electrophoresis chamber** and covered with TAE buffer (Tris-acetate-EDTA), which was also used to make the gel (**Figure 10.6**). At this point, we are ready to use a micropipette to transfer our DNA samples into the wells.

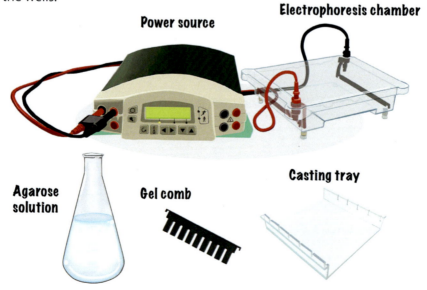

Figure 10.6: Equipment used in gel electrophoresis

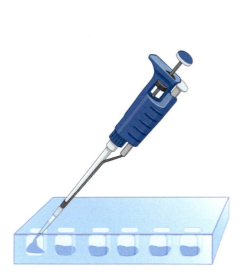

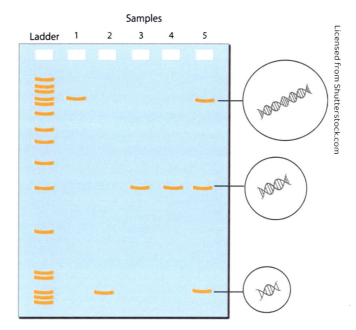

Figure 10.7: Pipetting DNA sample mixed with loading dye into the top of a gel in preparation for electrophoresis. The tip should not go to the bottom of the well, otherwise it can puncture the gel. Figure created with BioRender.com

Figure 10.8: A depiction of a gel showing that small pieces of DNA progress more quickly through the gel than larger pieces of DNA. The DNA ladder is also shown that allows size comparison.

Our DNA samples are clear liquids in small sample vials. Since we will be loading our samples into the gel underwater, we add a **loading dye** to make our samples blue, allowing us to see where our pipette tip is going as we place the DNA into the wells (**Figure 10.7**). The loading dye makes samples denser, ensuring they stay in the wells, and also allows us to track progress, preventing accidental loading of multiple samples into the same well. When loading the wells, also make sure to keep your tip away from the bottom of the well so that you do not puncture the gel below the well. Puncturing the gel may result in losing your sample. In addition to our samples, we will load a DNA ladder to help with analysis. A **DNA ladder** is a mixture of DNA fragments of known lengths, which serves as a reference to determine the size of the sample fragments by comparing our fragments to the ladder fragments.

Once the samples are loaded into the gel, we connect the electrophoresis chamber to a power source. The amount of voltage and amperage applied depends on the characteristics of our analysis. DNA is negatively charged, thanks in part to the phosphate groups that make up the backbone of the molecule, so the DNA will migrate towards the positive electrode. As the DNA fragments move through the gel, smaller fragments travel faster and further than larger ones that are more inhibited by the agarose, creating a pattern of bands. These bands can then be compared to the DNA ladder to determine the fragment sizes, helping us identify the genetic origin of the sample (**Figure 10.8**).

Since DNA is colorless, we must stain it to visualize the fragments. One common method is to mix a **pre-stain** into the gel before it solidifies (around 10uL). As the DNA moves through the gel, it picks up the stain, causing the bands to glow under specific wavelengths of light. Some stains require UV light, but many now require other types of light that are not harmful (like blue light). Originally, these stains were highly toxic, but now there are safer alternatives available, like the one we will use today. Regardless, it is essential to always wear proper protective equipment when handling these chemicals.

EXERCISE: LOOKING AT HOW GELS SEPARATE DNA FRAGMENTS

In this exercise we will continue to use the data from our EcoRI enzyme digest in the previous exercise. This time, we will depict what the restriction enzyme digests would look like if we were to run them on a gel to better understand how gels separate fragments by size. Below is an image of a gel with open boxes along the top that represent the sample wells. DNA will move from the top toward the bottom. We have added DNA ladder to well 1 and our three samples will go into wells 2-4. Draw the pattern of bands that would be produced from each of the restriction digests for suspects A, B, and C from the previous exercise (suspect A goes in sample well A, etc.). The bands from the DNA ladder have been filled in for you already, so that you can use them as a guide to fill out the rest of the gel.

QUESTIONS

1. Did any of the suspects fit the DNA profile from the crime scene?

2. Could you ever end up with fragments that are the same size, but have a different sequence?

EXERCISE: LOAD AND RUN YOUR OWN GEL TO TEST THE MYSTERY MEAT

Now that you understand how gels separate the fragments of DNA we made with the restriction enzyme digest, we can proceed to run our own gel to determine whether the suspect meat is from a whale or fish. All the equipment we will need, as seen in Figure 10.5, is set up at the front of the room. The instructor has prepared the gels ahead of time, so we only need to load our samples.

We will be including our meat sample DNA along with DNA from a swordfish and DNA from a minke whale. All three samples have been subjected to two different restriction enzyme digests (we will call these enzymes A and B for simplicity). This helps ensure that we do not get identical bands in error, giving us greater confidence in determining whether our mystery meat is whale or fish. This means that we will have two wells for each sample. We will also include a DNA ladder in the last well to help us identify the size of the bands we see from our samples. The samples to be loaded into our gel are summarized in Table 10.1.

Table 10.1: The list of samples that will be put into the corresponding gel wells

Gel Lane	PCR Tube	DNA Sample
1	Tube A	Meat sample cut with enzyme A
2	Tube B	Meat sample cut with enzyme B
3	Tube C	Swordfish DNA cut with enzyme A
4	Tube D	Swordfish DNA cut with enzyme B
5	Tube E	Minke whale DNA cut with enzyme A
6	Tube F	Minke whale DNA cut with enzyme B
7	Tube G	DNA Ladder for analysis

PROCEDURE:

1. If the instructor has not already done so, load your gel into the electrophoresis chamber. The chamber should have enough TAE buffer to cover the top of the gel. If the buffer does not cover the top of the gel, ask the instructor to add more to the chamber.

2. Take note of the DNA samples. They are in PCR tubes that are held together in a strip format. Loading dye has already been mixed with the sample to speed up the process. After the loading dye was added, the sample was spun in a small tabletop centrifuge to ensure all the liquid is at the bottom of the PCR tube. We are working with very small volumes of liquid, so we need to ensure it is where we need it.

Procedure continued onto next page.....

3. Remember how to use a micropipette (refer to **Figure 10.9**). Carefully insert your micropipette into the sample of the first PCR tube and extract the sample. We will be placing the first sample in the first well, the second sample in the second well, and so on.

4. Now, carefully position your pipette above the well you intend to load. Avoid touching the gel to prevent damaging it or disturbing any other samples already in the gel. Dip the tip of your pipette into the well without touching the bottom to avoid poking a hole in the gel or blocking the sample from leaving the tip (**Figure 10.10**). Once your pipette tip is in place, gently push down on the plunger to release the blue mixture of DNA and loading dye into the well. Keep your thumb pressing down on the plunger until all the sample is dispensed and you have removed the empty tip from the buffer.

5. Your instructor will now make sure that the DNA laddeer is loaded with your samples, set the power supply to the appropriate settings (around 160 V), and start the power. The power will remain on until the loading dye gets near the bottom of the gel. Observe the gels while they are running.

6. Once done, the instructor will cut the power and begin the next steps to visualize the gel.

For volumes between 1-20 µl:

1. First adjust the pipette to desired volume

2. Take a tip from the tip box with the micropipette

3. Press on the micropipette plunger to the first level

4. Place the micropipette in the solution

5. Release the micropipette head

6. Take the pipette out of the solution

7. Insert the tip into the delivery vessel

8. Press on the micropipette plunger to the second level

9. Eject the tip into the waste container

Figure 10.9: Instructions for how to use a micropipette.

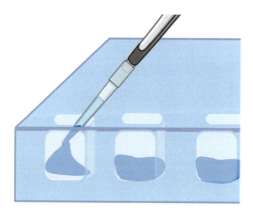

Figure 10.10: Exercise caution when pipetting sample into a gel. Keep your tip only about half way down into the gel to avoid poking a hole in the bottom. Figure created with BioRender.com.

10.4 Interpreting our gel band patterns: Illegal whale or fish?

Restriction Fragment Length Polymorphism (RFLP) is a genetic fingerprinting technique used in gel analysis. It involves analyzing the variations in the length of DNA fragments produced by the digestion of DNA with specific restriction enzymes. The term "fingerprinting technique" comes from the fact that the areas used for these types of analyses are variable enough to produce reliably different patterns of bands for DNA from different individuals (or species). When running a gel with DNA from several individuals or species, we expect to see differences in the lengths of DNA fragments cut by each restriction enzyme and potentially different numbers of fragments. This produces a unique banding pattern in the gel for each individual or species in the analysis.

In our analysis, we use two different restriction enzymes to act on three DNA samples: the suspected whale meat, DNA from a swordfish, and DNA from a minke whale. By comparing the banding patterns produced by both enzymes for each sample, we increase the reliability of our conclusions. Using two enzymes allows us to confirm that the observed patterns are not due to an anomaly with a single enzyme's cut site. This redundancy helps ensure that the results are accurate and that the genetic differences are correctly identified. To aid in this comparison of banding patterns, we include a DNA ladder in the last well of our gel. The DNA ladder serves as a reference, allowing us to determine the sizes of the bands in our samples (**Figure 10.11**).

Before we can interpret the gels, they must first be taken from the electrophoresis chamber to a light source called a **transilluminator** that will allow us to see the bands. Remember, the bands have a pre-stain that causes them to fluoresce.

PROCEDURE:

1. Your instructor will take the gel from the electrophoresis chamber to the transilluminator (**Figure 10.12**).

2. Each gel will be placed on the transilluminator and the blue light will be turned on. It is important for the room lights to be dimmed or turned off to fully see the DNA bands glowing.

3. Take a picture of your gel to analyze in the steps that will continue on the next page. Make sure to have the gel fill the screen of your phone.

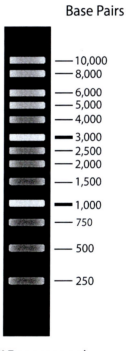

Base Pairs

— 10,000
— 8,000
— 6,000
— 5,000
— 4,000
— 3,000
— 2,500
— 2,000
— 1,500
— 1,000
— 750
— 500
— 250

1% TAE agarose gel

Figure 10.11: The DNA ladder key showing the sizes of the DNA bands produced by the ladder in a gel.

Figure 10.12: The transilluminator. Created with BioRender.com

Now that you have a picture of your gel, how do you interpret what you see? Look for unique banding patterns for each DNA sample and compare them against the DNA ladder. The differences in the patterns will help you determine whether the suspected whale meat matches the DNA of a whale or if it is from a different species, like a swordfish. Refer to the example (shown in **Figure 10.13**), which closely parallels our lab activity. In the example, a rectangle is drawn around the sample that matches the banding pattern of the suspected whale meat. This example will help you visualize how to compare and interpret your own results. Some specific steps are outlined below to help you make this comparison, along with a related example of how to interpret a gel.

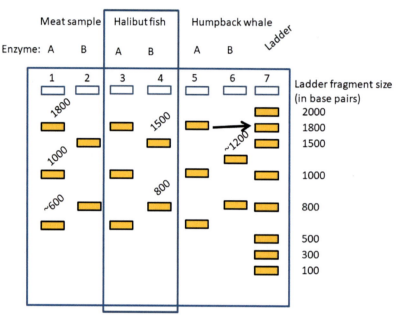

There are two wells per DNA sample. Well A has the sample cut by restriction enzyme A and well B has DNA cut by restriction enzyme B

Figure 10.13: An example gel similar to the one you will work with that shows the banding patterns and demonstrates how to interpret the bands against the DNA ladder.

INTERPRETATION STEPS:

1. Figure out approximately how large each DNA fragment is by matching the bands of DNA against the ladder. Make a rough drawing of your gel results in the **Figure 10.14**. In **Table 10.2**, record the approximate sizes of the bands in each lane.

2. Compare the banding patterns in the wells containing the DNA samples from the suspected whale meat, swordfish, and minke whale.

3. Identify which sample has bands that match the banding pattern you see in the wells containing the suspected whale meat. Look for identical or very similar band sizes and patterns in both enzyme digests.

Figure 10.14: The banding pattern from your own analysis of the suspected whale meat

Table 10.2: The size of DNA bands from the restriction enzyme digest and gel electrophoresis.

Lane	DNA Sample	DNA Fragment length (in bp)
1	Meat sample cut with enzyne A	
2	Meat sample cut with enzyme B	
3	Swordfish cut with enzyme A	
4	Swordfish cut with enzyme B	
5	Minke whale cut with enzyme 1	
6	Minke whale cut with enzyme 2	
7	Ladder	No need to put info here

ANALYSIS OF THE RESULTS

1. Based on the banding patterns, was the unknown meat from a whale or fish?

2. Could those DNA samples have been distinguished from each other if just Enzyme A had been used? Explain why or why not.

Chapter 10 Post Lab Questions

1. Explain what a restriction enzyme is in your own words.

2. The EcoRV is a restriction enzyme that recognizes the specific DNA sequence 5'-GATATC-3' and cuts between the T and A nucleotides, producing blunt ends. Demonstrate where the enzyme would cut on the sequence below. Make sure to label the size of each fragment produced.

5'- AGTCCGATATCGTACGTAGCTAGATATCGGCTAGCTAGGATATCTTACGTAGCCGTAAGTCCG -3'

3. The EcoRV enzyme produces blunt ends. What does this mean?

4. Gel electrophoresis takes advantage of the fact that the DNA has an overall negative charge on the molecule. Where does the negative charge come from?

5. The DNA is generally not visible to the naked eye. How did we go about coloring the DNA solution when we loaded the samples? How did we make the DNA visible as bands in our gel?

6. What is the purpose of loading DNA ladder in one of the sample wells?

7. In what direction do the DNA samples travel in the gel when we turn the power on and why?

8. What determines how quickly a particular fragment of DNA travels through a gel?

9. Why did we use two different restriction enzymes to analyze our three samples instead of just one enzyme?

10. What do we need to be careful not to do when loading our gel with samples using the micropipette?

Notes

Notes